T0240128

Glaciology and Glacial Geomorphology

Wilfried Hagg

Glaciology and Glacial Geomorphology

 Springer

Wilfried Hagg
Fakultät für Geoinformation
Hochschule für angewandte Wissenschaften
München, Germany

ISBN 978-3-662-64716-5   ISBN 978-3-662-64714-1   (eBook)
https://doi.org/10.1007/978-3-662-64714-1

Acknowledgements

As I write these lines, it is now almost 20 years to the day since I took up my doctoral position at the Bavarian Academy of Sciences and Humanities. Since then, I have found a workplace, a coffee round, and a lot of discussion and expertise around the topic of glaciers in this house. The biggest thanks go to my colleagues from the Geodesy and Glaciology group for this pleasant and friendly environment and to the members of the running group for the inspiring conversations in the English Garden.

While working on this book, I also enjoyed excellent working conditions in the Department of Geography at Ludwig-Maximilians-University, where parts of the manuscript were written at the original desk of Erich von Drygalski, as well as in the Department of Geoinformatics at the Munich University of Applied Sciences.

I would also like to thank the German Research Foundation for its multiple grants. Without this support, it would probably not have been 20 years that I was able to deal with this beautiful and multifaceted topic.

A big Dankeschön goes to Heidi Escher-Vetter, Ulrich Münzer and Simon Feigl for their valuable comments on the German first edition.

Munich, Germany, May 2020

Contents

Introduction and History of Research

Contents

© The Author(s), under exclusive license to Springer-Verlag
GmbH, DE, part of Springer Nature 2022
W. Hagg, *Glaciology and Glacial Geomorphology*,
https://doi.org/10.1007/978-3-662-64714-1_1

1

Overview
After a brief description of how glaciers are perceived by the public today, the object of research, tasks and methods of the two scientific disciplines mentioned in the title of the book are described. A definition of terms and important site designations on, in and under glaciers is followed by an outline of the history of research – from the first descriptions in the sixteenth century to the dispute between plutonists and neptunists to the technological progress that began in the twentieth century and continues to influence developments today.

The term "glacier" evokes different images in us. In an experiment with first-semester students, the terms "cold", "ice age" and "disappear" were most frequently mentioned during free association. It is hardly surprising that people think of cold when they think of frozen water, although glaciers can sometimes be exposed to double-digit temperatures in summer. The mention of ice age in the above-mentioned experiment may have something to do with Munich as a place of study. Here, the fact that glaciers reached as far as our front door 20,000 years ago may well have been anchored in the collective consciousness through school lessons alone. The fact that today they are following a contrary trend and are even threatening to disappear is probably associated with glaciers worldwide. The rapid changes of recent years have led to comparative images of glaciers becoming almost emblematic of climate change. They show the consequences of global warming particularly clearly and make them comprehensible and understandable for everyone.

Another association is purity. Glaciers seem to exist far away from human pollution and are therefore considered particularly pure – an aspect that is often taken up in the marketing of water. Glaciers exert a fascination because they are beautiful and dangerous at the same time. The ice sometimes glows downright turquoise in the sunlight and, like a diamond, gives the final touch to the beauty of a mountain range. At the same time, crevasses arouse primal fears in mountaineers. But not very long ago, the glaciers themselves also struck fear into mountain dwellers when they advanced and seemed to herald a new ice age. This image has changed considerably, and at present one feels rather sorry for the languishing and seemingly moribund patients. Where in the 19th century there were still processions of supplication against glacier advances, as for example in the Valais community of Fiesch, today there are organized prayers against the glacier recession. When Okjökull became the first glacier in Iceland to fall victim to climate change in the summer of 2019, it was a media event recorded around the world.

Consequently, the main perception of glaciers at present is their importance as climate indicators. However, they are also natural reservoirs that can redistribute water on different temporal scales and significantly influence the availability of this valuable resource. In some arid areas, where food production is only possible through irrigation, glacial meltwater is the most important or even the only source of water during the main vegetation period. As hydrological reservoirs, glaciers are seen by the public mainly for their contribution to sea-level rise, which they provide through their melting. This is admittedly of global interest and may change our planet in the long term, especially if the two ice sheets in Greenland and Antarctica are included.

Somewhat less well known is the fact that glaciers are active landscape shapers and have in the past shaped not only high mountain relief but also large areas of North American and Central and Northern European lowlands. Anyone wishing to understand the formation of these glacial land surfaces cannot avoid considering the processes and effects of glacier erosion and deposition. In the following, the object of research, tasks and methods of glaciology and glacial geomorphology as scientific disciplines will be described and their historical development briefly illuminated.

1.1 **Object of Research**

Glaciology is the study of glaciers, or more generally ice and all phenomena that are connected to it.

Today, glaciology is an interdisciplinary field of research involving geodesists, cartographers, meteorologists, physicists and geophysicists, crystallographers, hydrologists and geographers, among others. The most important tasks are research into the relationships between climate and glaciers, the observation and interpretation of glacier fluctuations, the quantification of the hydrological consequences of glacier retreat and the study of glacial natural hazards.

The most important working techniques are observation and measurement in the field, remote sensing by means of aerial and satellite photographs, simulation (e.g. of ice movement or mass balance) with mathematical models and laboratory analyses (e.g. of firn and ice cores).

Glacial geomorphology, as the study of glacial erosion and sedimentation, is a subfield of landform science, i.e. geomorphology, and thus in turn a discipline of physical geography. However, there is overlap with geology, which also studies the formation of glacial and other landforms as processes of exogenic dynamics. Glacial geomorphology is concerned with studying the processes of glacial erosion and deposition, clarifying the formation of glacial landscapes, and reconstructing glacial and climatic history. Glacial geomorphologists also make use of the above-mentioned spectrum of methods; additional working methods are dating of glacial deposits, which in turn can be subdivided into field and laboratory methods.

This textbook deals with glaciology and glacial geomorphology in the sense just outlined; it is intended for students of geography and related geosciences, interested laypersons, mountaineers and nature lovers. The central topic is glaciers, their dynamics and their influence on the formation of the earth's surface.

> ❯ A glacier is defined as an actively moving mass of snow, firn and glacial ice. In addition, meltwater and entrained rock are also components of a glacier.

The decisive criterion is the active movement by which the glacier distinguishes itself from other ice deposits. Since this common definition does not require a minimum amount of ice movement, and there is a fluent transition towards zero with the current shrinkage of glaciers, it is often difficult to say at what point a glacier ceases to be one. A minimum size, as often demanded and also applied in some inventories, is arbitrary and physically unjustifiable. In steep relief, even very

1

small ice patches can still move and thus meet the criteria for a glacier. Depending on the application, however, a demarcation must sometimes be made because, for example, not every smallest ice patch can be taken into account in global surveys of ice reserves. For smaller study areas, however, it may be most practicable to treat each glacier as such until the last ice has disappeared.

The German word "Gletscher" first appears in the Swiss chronicle of P. Etterlin from 1507 and is derived from *glacies* (Latin for "ice"). An overview of the term in some of the languages and dialects in which it occurs is given in ◘ Table 1.1.

There is a multitude of terms describing locations and positional relationships on, in, under and around the glacier. Since they are part of the linguistic tools of the trade if one wants to talk or write about glaciers in a well-founded manner, they should be mentioned at this point. If one approaches a glacier from the valley, one first walks over the **glacier forefield**. This is the name given to the area in front of the glacier (**proglacial**), which is usually not ice-free for very long. The lowest end of the glacier is called the **glacier front**. This is also the location of the **glacier gate**, a more

◘ **Table 1.1**   The term "glacier" in different languages and dialects

Language/Dialect	Designation
German	Glacier
German/Tyrolean	Ferner (from Firn, Old High German: alt), Kahr (Tauern)
German/Pinzgauan	Kees (Old High German: ice)
German/Carinthian	Kess, Käss
German/Swiss-German	Firn, Firre
Romanic	Glatsch, Glatscheret, Vadret, Vedrett
French	Glacier, Biegno (Valais), Serneille (Pyrenees)
Italian	Ghiacciaio, Vedretto, Ruize (Piedmont), Rosa (Valle d'Aosta)
Slovenian	Ledenik
English	Glacier
Spanish	Glaciar, Helero, Nevero
Spanish/chilean	Ventisquero
Swedish	Glaciär
Norwegian	Brae, Isbrae, Breen
Lappish	Jegna
Icelandic	Jökull
Greenlandic	Sermek (land ice), Sermerssuak (inland ice)
Russian	Lednik
Chinese	Bīngchuān, Bīnghé ("ice river")

◻ Fig. 1.1 Glacier terminus with glacier gate in the Ötztal. The position of the photographer is proglacial, i.e. he is standing in the glacier forefield. (Photo: W. Hagg)

or less large opening from which a meltwater stream exits the glacier (◻ Fig. 1.1). Processes that take place at the front of the glacier are called **frontal**, those at the sides of the glacier **lateral**.

Transitional areas are sometimes called **latero-frontal**. The lower part of a glacier is typically elongated and oriented along the course of the valley; it is called the **glacier tongue**. Anything that lies or takes place on the glacier is called **supraglacial**. Corresponding locations in the glacier are called **englacial** or **intraglacial**, while deposits or processes beneath the glacier are called **subglacial**. The **glacier bed** is the subsurface below of the ice, which may consist of solid or loose rock. The lowest metres of the glacier may contain rock fragments from the glacier bed; they are referred to as the **basal transport zone**.

1.2 History of Research

The oldest description of glaciers appears in the sixteenth century in the *Cosmographia* (Münster 1544–1628), the first scientific description of the world in German. Among other things, the author recounts how in 1546 he visited the Rhone glacier on horseback, "one crossbow shot wide". He found it threatening (Münster 1628), partly because a piece of ice the size of a house had fallen from it. Glacial ice was considered so cold that a piece the size of an egg was enough to make a jug of wine "grimly cold". These accounts prove that the phenomenon of glaciers was frightening to people at the time. Josias Simler, a Swiss theologian and historian, published the first comprehensive work on the Alps in Latin in 1574. He still adhered to the idea, naive from today's point of view, of the Roman naturalist Pliny the Elder, according to which rock crystal is formed from heavily frozen ice. As the Alpine glaciers advanced far towards valleys and villages during the Little Ice Age (▶ Chap. 8), historical sources of devastation began to accumulate from the seventeenth century onwards. In the verses of Hans Rudolf Räbmann ("*Stosst vor ihm weg das Erderich, Böum, Heuser, Felsen wunderlich*"), the author describes the destructive power of the glaciers in 1606. At the end of the seventeenth century, with the *Abhandlungen von den isländischen Eisbergen* (Vidalin 1695), reports on glaciers outside the Alps appear for the first time, but they are still mainly descriptive in nature.

1

A scientific study of glaciers began in the eighteenth century. Scheuchzer (1706–1708) already describes the stratification of firn and ice, but still explains glacier movement with the expansion of freezing water in crevasses. Bordier (1773) recognizes the plastic property of ice and provides explanations for the formation of moraines. On the basis of observations and simple measurements on the Grindelwald Glacier, some of which were made by a shepherd boy, Kuhn (1787) writes a highly regarded *Versuch über den Mechanismus der Gletscher*. In *Nachrichten von den Eisbergen in Tyrol,* Walcher (1773) for the first time makes climatic causes responsible for glacier fluctuations; until then they had been explained by God-ordained cycles and biblical figures. Numerous beautiful drawings, but also scientific findings, for example about the movement of ice due to pressure, are due to H.B. de Saussure (1779–1786) and his work *Voyage dans les Alpes*.

In the eighteenth century, glacial forms in the Alpine foothills were still misinterpreted. Kettle holes, for example, were mistaken for volcanic craters. The greatest mystery, however, was posed by so-called erratics. These sometimes huge boulders of non-local rock are so large that they cannot have reached their place of discovery by currently observable transport processes (◘ Fig. 1.2). The **Plutonist** school tried to explain the world of the time in terms of forces from the Earth's interior, and considered erratic boulders to be volcanic bombs. The **Neptunists**, on the other hand, blamed water for transport. Catastrophic tidal waves, for example from a glacial lake outburst, could have transported the erratic blocks. Another possibility would be a deluge, which allowed icebergs with enclosed boulders to be transported from the far north. These boulders are then said to have melted out and sunk to the seabed of the time as so-called dropstones.

The nineteenth century is considered the classic era of glacier research. The first halfway usable cartographic representations of glaciers were created through large surveying campaigns. Lieutenant Josef Naus, for example, surveyed the glaciers on the Zugspitzplatt on behalf of the Royal Bavarian Topographical Bureau (◘ Fig. 1.3) and climbed Germany's highest mountain for the first time with his surveyor's assistant Maier and the mountain guide Deuschl.

As early as 1824, the Danish geologist Jens Esmark saw the Norwegian fjords as the result of glacial erosion, propagated the existence of worldwide ice ages and held changes in the Earth's orbit responsible for their formation. However, it was

◘ **Fig. 1.2** The "Old Swede", a granite boulder weighing 217 t from Småland, which was transported to Hamburg by the Skandinavian Ice Sheet and was only dredged from the Elbe navigation channel in 1999 (Vinx 1999). (Photo: Corinna Grave)

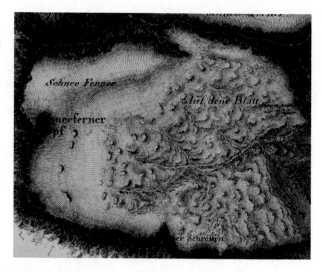

❑ **Fig. 1.3** Detail of the position sheet "Zug-Spitz" (No. 888). Survey (1820) and drawing (1826) by Lieutenant Naus. (Historical map: Bavarian Survey Administration; Creative Commons license CC BY-ND)

Louis Agassiz who decisively developed the ice age theory in 1836–1837, even though he adopted basic ideas from colleagues such as Esmark, Venetz, Charpentier or Schimper. The latter is also responsible for the term *"Eiszeit"* (Schimper 1837), which was later translated from German into all languages. But Agassiz also had a hard time in science at first, although he had collected many arguments before he presented his thoughts in a lecture. Much time was to pass before his ice age theory was generally accepted. It was not until 1875 that Otto Martin Torell was able to prove, on the basis of traces of ice in the Rüdersdorf limestone mountains, that northern Germany had been covered by inland ice during the Ice Age. Only now was the iceberg drift theory of the Neptunists finally disproved.

While in the first half of the nineteenth century glacier research was focused on the Western Alps, the brothers Schlagintweit and Schlagintweit (1850) added the first work from the Eastern Alps. These were glaciological, geological, meteorological and botanical studies in the Ötztal and in the area of the Grossglockner, followed by a famous journey to India and High Asia from 1854 to 1857, sponsored by Alexander von Humboldt (Schlagintweit 1869–1872). During this period, glaciological studies accumulated, for example the theory of glacial movement by melting and refreezing (Tyndall and Huxley 1857) or the formation of ice ages by variations in the Earth's orbital parameters (Croll 1864). The first mass balance measurements were made on the Rhone Glacier from 1874 onwards (Mercanton 1916), and in the Eastern Alps glacier observations were coordinated by the German and Austrian Alpine Clubs from 1888 onwards (Richter 1893). In order to systematise the increasing number of surveys, the *Commission Internationale des Glaciers* was founded in 1894. The first glacier maps from regions outside the Alps were also produced at this time, for example from Norway (Sexe 1864), the Caucasus (Merzbacher 1901), Kilimanjaro (Kraus and Meyer 1900), New Zealand (Douglas and Harper 1893) or Mt. Rainier (Russell 1898). However, these maps still contained hardly any evaluable elevation information except for a few summit heights. At the same time, Sebastian Finsterwalder perfected the surveying method

1

of photogrammetry, by means of which he produced the first precise map of an entire glacier at a scale of 1:10,000 and contour lines at intervals of 10 m at Vernagtferner in Tyrol in 1889. At the same time, Finsterwalder (1897) developed his kinematic-geometric theory of motion there, which, although it does not explain ice movement, still describes it correctly (▶ Chap. 3).

In the course of the first international polar year in 1882/1883, polar research was intensified, in which researchers from the German Empire also participated and which initially focused on the North Polar region. The first German Antarctic expedition took place from 1901 to 1903 under the leadership of the geographer Erich von Drygalski. Thirty-two participants wintered on a research ship, enclosed by sea ice. They collected so much data, including glaciological measurements, that the analysis took three decades and the results were published in over 20 volumes (Lüdecke 1995).

Glacial geomorphology also experienced a significant boost during these years. After Albrecht Penck won a prize at the University of Munich in 1880 for the explanation of glacial phenomena in the Bavarian Alpine foothills and the Bavarian Alps, he worked on this topic for 20 years until he published the standard work *Die Alpen im Eiszeitalter* together with Eduard Brückner in 1909 (Penck and Brückner 1901–1909). Penck was the first to prove the multiple glaciation of the Alps with field evidence. He succeeded in correlating meltwater sediments in the foreland with moraines in the mountains (◻ Fig. 1.4).

Penck's research was pioneering for decades; only much later were similar works from extra-Alpine regions published, for example by Wolstedt (1929) in northern Germany or by Gerasimov and Markov (1939) in the USSR.

In the twentieth century, glacier research was driven primarily by advances in aviation and technology. Aerial photographs have made aerial photogrammetry possible since 1923, thus simplifying high-mountain cartography. In the second half of the century, satellite images were added, which, in contrast to aerial photographs, provide greater spatial coverage and high temporal resolution (repetition rate 1–16 days). They are also often equipped with multispectral sensors, allowing

◻ **Fig. 1.4**   On these consolidated gravels south of Munich Albrecht Penck recognized the deposits of three different glaciations. (Photo: W. Hagg)

automatic or at least semi-automatic glacier delineation (Bhambri and Bolch 2009). Spatial resolution improved steadily, from 80 m with Landsat MSS (1975) to 61 cm with Quickbird (2001). Modern software can perform more and more steps automatically, for example, control point generation and image orientation in digital photogrammetry. In this way, the time required for evaluation can be greatly minimized. This method was used, for example, to create the Austrian Glacier Inventory 1998 (Kuhn et al. 2008), in which 896 glaciers with a total area of 471 km^2 were catalogued.

The digital processing of raster and vector data has significantly accelerated many work steps in all geosciences, and the use of Geographic Information Systems has opened up new possibilities for data analysis and geovisualization.

Another modern measurement technique is laser scanning, in which the earth's surface is scanned by means of electromagnetic waves (Baltsavias et al. 2001) and which can provide very accurate elevation models when used terrestrially, by aircraft or by satellite (◘ Fig. 1.5).

But it is not only the glacier surfaces that can be recorded with increasing precision; radar measurements also make it possible to look into layers of snow, firn and ice. Even before the Second World War, pilots reported unrealistically low flight altitudes over inland ice. The explanation that the radar waves of the altimeter penetrated the ice was quickly found; since 1929 this effect has been used for ice thickness measurements. A radar antenna sends pulses into the ice, which are reflected by the bedrock and recorded again by a receiver. Because the propagation of the radar waves in the ice is constant, the ice thickness can be calculated from the propagation time of the signal. ◘ Figure 1.6 shows an exemplary evaluation on the Zugspitzplatt, the purple shades indicate the reflection, the left x-axis the propagation time of the signal and the right one the depth of the bedrock. The strong reflection at approx. 74 m travel distance (x-axis) is an artefact, here a piece of metal was probably just below the surface.

Synthetic aperture radar (SAR) provides two-dimensional images that look similar to photographic images. As an active remote sensing system, however, it has the immense advantage over optical methods that it can penetrate clouds and can also be used at night. In the case of TerraSAR-X, a German satellite that has been in orbit since 2007, calculations of elevation models with a ground resolution of only 5 m are possible during two overflights with a slightly offset orbit. Since June 2010, the TanDEM-X satellite has been in formation flight with TerraSAR-X, enabling the creation of a global terrain model with unprecedented accuracy (up to 1 m ground resolution, height accuracy <2 m). On a local scale, the use of drones has also become increasingly important for glacier monitoring in recent years. In the foreseeable future, the challenge will no longer lie in geodetic surveying and observation of glaciers, but much more in researching ice dynamics and the climate-glacier relationship, as well as in estimating future changes and their consequences.

Fig. 1.5   Perspective 3-D view of the Schneeferner on the Zugspitzplatt. The section corresponds approximately to the left half of ▪ Fig. 1.3. A historical map (Finsterwalder 1892) and aerial orthophotos of the remaining glacier remnants (framed in red) of the Southern (left) and Northern Schneeferner (right) from 2018 were draped over an elevation model from laser scanning data from 2006. (source: LDBV). The yellow line on the Northern Schneeferner shows the location of the radar profile of ▪ Fig. 1.6

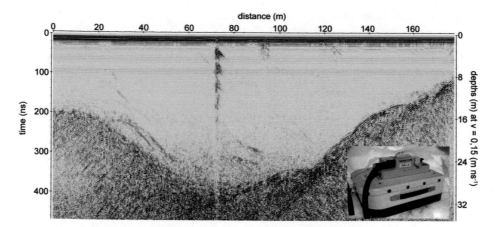

distance (m)

Fig. 1.6 Radargram of a transect over the Northern Schneeferner in 2007 and the radar antenna used (small photo). The position of the measured profile is marked in ◻ Fig. 1.5

References

Baltsavias EP, Favey E, Bauder A, Boesch H, Pateraki M (2001) Digital surface modelling by airborne laser scanning and digital photogrammetry for glacier monitoring. Photogramm Rec 17(98):243–273

Bhambri R, Bolch T (2009) Glacier mapping: a review with special reference to the Indian Himalayas. Prog Phys Geogr 33(5):672–704

Bordier AC (1773) Voyage pittoresque aux glaciers de Savoye 1772. L.A. Caille, Genf

Croll J (1864) On the physical cause of the change of climate during geological epochs. Philos Mag 28(187):121–137

de Saussure HB (1779–1786) Voyage dans les Alpes. Fauche, Genève

Douglas C, Harper AP (1893) Topographical plan of Waiho country. Appendices to the Journals of the House of Representatives, 1894, C-1. https://teara.govt.nz/en/zoomify/10751/map-of-the-franz-josef-glacier. Accessed 12 July 2020

Esmark J (1824) Bidrag til vor Jordklodes Historie. Mag Naturvidenskaberne Aarg. 2 1:28–49. Serie III

Finsterwalder S (1892) Zugspitze. Karte 1:10000. Bearbeitet im Topographischen Bureau des k. b. Generalstabes

Finsterwalder S (1897) Der Vernagtferner. Z.D.Ö.A.V. 60:143–156

Gerasimov IP, Markov KK (1939) Die Eiszeit auf dem Territorium der UdSSR. Die physisch-geographischen Bedingungen der Eiszeit.- Akademie d. Wiss. UdSSR. Arb. des Instituts f. Geogr. Lfg 33, Moskau-Leningrad 1939, 196 Abb. i. Text u. a. Taf. [russ., engl. Zus.fass]

Kraus P, Meyer H (1900) Spezialkarte des Kilimandjaro: nach den neuesten Aufnahmen von Prof. Dr. Hans Meyer und mit Benutzung von Messungen, Entwürfen und Skizzen von Hauptmann Johannes, Dr. Carl Lent, Oberst v. Trotha, Graf Wickenburg, Dr. A. Widenmann u. a. Dietrich Reimer, Berlin

Kuhn BF (1787) Versuch über den Mechanismus der Gletscher. Magazin über die Naturkunde Helvetiens 1:119–136

Kuhn M, Lambrecht A, Abermann J, Patzelt G, Groß G (2008) Die österreichischen Gletscher 1998 und 1969, Flächen und Volumenänderungen. Verlag der Österreichischen Akademie der Wissenschaften, Wien

Lüdecke C (1995) Die deutsche Polarforschung seit der Jahrhundertwende und der Einfluss Erich von Drygalskis = German polar research since the turn of the century and the influence of Erich von

Drygalski, Berichte zur Polarforschung (Reports on Polar Research), vol 158. Alfred Wegener Institute for Polar and Marine Research, Bremerhaven. https://doi.org/10.2312/BzP_0158_1995

Mercanton PL (1916) Vermessungen am Rhonegletscher 1874–1915. Kommissions-Verlag von Georg & Co, Basel

Merzbacher G (1901) Aus den Hochregionen des Kaukasus. Wanderungen, Erlebnisse, Beobachtungen, vol 2. Duncker & Humblot, Leipzig

Münster S (1544–1628) Cosmographia. Heinrich Petri, Basel

Münster S (1628) Cosmographia. Faksimile-Druck von 1984. Antiqua, Lindau

Penck A, Brückner E (1901–1909) Die Alpen Im Eiszeitalter, vol 3. Tauchnitz, Leipzig

Richter E (1893) Bericht über die Schwankungen der Gletscher der Ostalpen 1888–1892. Z Dtsch Österr Alpenvereins 24:473–485

Russell IC (1898) Glaciers of Mount Rainier. USGS 18th Ann. Rept. 1896–1897, vol 2, pp S 355–S 415

Scheuchzer JJ (1706–1708) Beschreibung der Naturgeschichte des Schweizerlandes, vol 7. Heidegger und Compagnie, Zürich

Schimper KF (1837) Über die Eiszeit. Actes Soc Helv Natur Neuchâtel 22:38–51

Schlagintweit H, Schlagintweit A (1850) Untersuchungen über die Physicalische Geographie der Alpen. Johann Ambrosius Barth, Leipzig

Sexe SA (1864) Om Sneebraeen Folgefond. University Programme, Christiania

Tyndall J, Huxley TH (1857) On the structure and motions of glaciers. Philos Trans R Soc Lond 14:327–346

Vidalin P (1695) Dissertationcula de montibus Islandiae chrysta. In: Vidalin P (Übers.) (1754): Abhandlung von den isländischen Eisbergen. Hamburgisches Magazin 13:9–27

Vinx R (1999) Der Elbfindling von Hamburg-Övelgönne. Geschiebekunde aktuell 15(4):111–112

von Schlagintweit H (1869–1872) Reisen in Indien und Hochasien, Bd 3. Hermann Costenoble, Jena

Walcher J (1773) Nachrichten von den Eisbergen in Tyrol. Kurzböcken, Wien

Wohlstedt P (1929) Das Eiszeitalter. Grundlinien einer Geologie des Diluviums. Enke, Stuttgart

Origin of Glaciers

Contents

W. Hagg, *Glaciology and Glacial Geomorphology*,
https://doi.org/10.1007/978-3-662-64714-1_2

2

> **Overview**
> For glaciers to form in the high mountains, topographical and climatic conditions must be met. The different types of snow boundaries play a decisive role here. Glacial ice, a material with special physical properties, is formed from snow via various transformation processes. The temperature of the ice is of central importance for the movement of the glaciers and the shaping of the bedrock.

2.1 Preconditions for Glacier Formation

To put it simply, glaciers form where more snow falls than melts on average over the year, i.e. at locations where the warm season is relatively cool. Such conditions are found in regions close to the poles and in high mountain ranges. If we look at the matter a little more closely, it is not only necessary for snow to fall, but it must also be topographically possible for it to remain and accumulate. So in addition to climate, relief also plays a role: glaciers can only form where there are flattening zones and where large amounts of snow can accumulate over the years. Glaciers cannot form in steep relief, even if the climatic conditions (snowfall > melt) are met. The build-up and depletion of the snow cover is not only caused by snowfall and melting, but also by other processes such as avalanches or wind drifts.

❯ Glaciers are formed where, on average, more snow is deposited than is lost through melting and other processes.

These conditions are met above the **climatic snow line.** This is the name given to the altitude, averaged over several years, at which snow exists all year round. ◻ Figure 2.1 shows the planetary change of the climatic snow line from 80° northern to 70° southern latitude.

Even though the mean annual temperatures have increased since the measurement period on which ◻ Fig. 2.1 is based, some regularities are discernible which still apply today. In the northernmost areas, the snow line is at about 500 m a.s.l., and from 70° N it rises towards the equator due to the increasing temperatures. In the latitudes of the Alps (about 46–47° N) it is at 3000 m a.s.l., and in the subtrop-

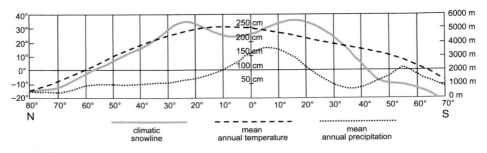

◻ **Fig. 2.1** Latitude-dependent variation of the climatic snow line. (Modified after de Martonne 1948)

ics (about 23° N and 18° S) it reaches its greatest heights of over 5000 m a.s.l. In between, in the inner tropics near the equator, it is somewhat lower at 4500 m a.s.l., which is due to the distribution of precipitation. In the subtropical high pressure belt it is hot and dry, which is the least favourable combination for permanent snow and explains the maximum values of the snow line. In the humid tropics near the equator, it is only marginally hotter, but has considerably more precipitation, and thus more snow at high elevations. This illustrates that the conditions for the formation and existence of glaciers are always controlled by two climatic factors, temperature and precipitation. In the southern hemisphere, because of the lower temperatures due to the lack of land masses ("water hemisphere") and the higher precipitation, the snow line falls more steeply and already reaches sea level at about the Tropic of Cancer. This is supported by the Antarctic Circumpolar Current, an ocean current that contains particularly cold water and climatically isolates Antarctica.

The course of the climatic snow line in ◘ Fig. 2.1 shows that it depends on the latitude. However, this is not the only influencing factor. Latitude has a decisive influence on radiation totals and mean temperatures, but precipitation in particular is still dependent on a variety of other control variables such as land-sea distribution, continentality, wind direction, and so on. Therefore, lines of equal height of the snow line do not always run parallel to the latitude. In western Alaska, for example, they are oriented north-south, because the precipitation-bearing winds come from the west and the snow amounts decrease rapidly inland.

While climatic snow lines always represent a regional average, **orographic snow lines** are based on a local scale. For this reason, small-scale differences resulting from relief and exposure must be taken into account here. On a mountain or mountain massif, for example, the orographic snow line may be lower on the windward side, i.e. the side facing precipitation, than on the leeward side. However, a strong influence of wind drift can also result in an opposite picture, when snow is blown out of windward slopes and accumulates in sheltered leeward situations. But exposure also affects solar radiation. On sunny slopes (in the northern hemisphere: south-facing), snow melts more quickly and the snow line is consequently higher than on shady slopes (◘ Fig. 2.2).

◘ **Fig. 2.2** Glaciers in the Suek Range, Kyrgyzstan. The fact that only north-facing valleys support glaciers illustrates the influence of orientation on the altitude of the climatic snowline. (Image taken by the Sentinel-2A satellite on 11 August 2019, © Copernicus 2019)

2

❯ On a glacier, the local snow line is called *equilibrium line.*

Due to the smoother surface of glaciers, it is usually more clearly defined than in the rough rocky terrain and, because of the cold content of the ice masses and the sinking cool air masses above them, is often considerably lower than the snow line in the rock surroundings.

For the sake of completeness, the **temporary snow line** should also be mentioned. This is not the highest position of the snow line in the course of the year, but the current lower limit of snow distribution. It is naturally subject to large seasonal and short-term fluctuations, whereby a descent due to snowfall can be completed much more quickly than an increase due to energy-intensive melting processes.

2.2 Processes Involved

2.2.1 Snowfall

In the natural temperature range of the atmosphere, water can occur in the three aggregate states solid, liquid and gaseous (◻ Fig. 2.3). During phase transitions, either energy must be expended (during sublimation, melting and evaporation), or energy is released (during desublimation, freezing and condensation). In meteorology, this so-called latent heat (*latens* for "hidden") is of great importance for energy conversions in the atmosphere.

Clouds consist of liquid water droplets or ice crystals. Because distilled water without impurities can be cooled to well below 0 °C, both aggregate states often exist simultaneously. If such a supercooled water droplet comes into contact with a solid suspended particle (aerosol), the latter acts as a freezing nucleus and the water immediately crystallises. Natural aerosols can be organic (e.g. pollen) and inorganic (dust, ash, salt) particles; combustion products are also introduced into

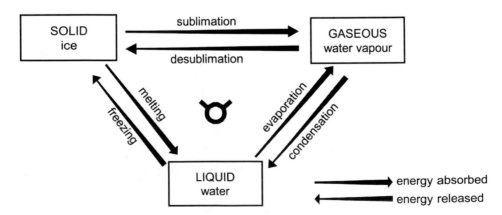

◻ **Fig. 2.3**   Aggregate states and phase transitions of water in the atmosphere

the atmosphere by humans. Layers of freezing water molecules are deposited on the surface of the ice particle. The ice crystals grow in a hexagonal structure, which was already recognized and described by Johannes Kepler in 1611. Depending on the humidity and temperature conditions in the cloud, however, different crystal shapes are formed. The Japanese snow researcher Nakaya (1954) classified these and presented them in a famous diagram, from which laws about their formation can be derived (▶ Excursus 2.1).

Individual snow crystals are at most 5 mm in size; the classical snowflake is an irregular agglomerate of many crystals that have been connected by supercooled droplets. For this to happen, the droplets must not be too severely supercooled so that they do not immediately freeze to the crystal, but there is enough time for another crystal to hit the same spot and become attached (Weischet 1991). Whenever snow crystals grow to such an extent that they can no longer be held in suspension by the updraft, snowfall begins.

Excursus 2.1: Snow Crystals

"Each snowflake is a letter from heaven." With this quote, Ukichiro Nakaya expressed that snow crystals can tell a lot about meteorological conditions in higher layers of the atmosphere. He photographed more than 3000 natural snow crystals on Hokkaido, which he divided into 40 categories. In the laboratory, he managed to recreate most of the shapes, thus establishing the link between atmospheric conditions and resulting shape. According to his results, the air temperature controls whether plates or prisms are formed (◘ Fig. 2.4). The complexity of the shape increases with humidity: At a slight supersaturation, simple columns (prisms) and plates are formed; at a stronger supersaturation, needles or complex plates called dendrites may form. Only these four basic types exist, the large variety of forms results from mixed and transitional forms. As conditions in the cloud change, crystals can begin their growth in one form and continue to grow in another (Furukawa 1997). This makes the "letter from heaven" longer and more interesting.

◘ **Fig. 2.4** The four basic types of snow crystals, from left to right: needles, columns, platelets, and a dendrite. (Photos by Kenneth Libbrecht; ▶ snowcrystals.com)

2

2.2.2 Snow Metamorphosis

Every child knows that there are different types of snow and that sticky snow is the most fun to play with. Even winter sports enthusiasts have their own terms such as "powder", "crud" or "crust". On the one hand, snow already differs while it is falling or freshly fallen, and on the other hand, the texture of a snowpack also changes over time. Various processes are involved in this (de Quervain 1963).

> As soon as snow reaches the earth's surface, a transformation process begins that is called **snow metamorphosis**. A distinction is made between **destructive** and **constructive metamorphosis**.

In the case of destructive metamorphosis, smaller and simpler end products are formed from larger and more complex initial products. The disintegration takes place by melting and refreezing or – in the case of **isothermal metamorphism** – mechanically and via the gas phase. The controlling factor in the latter case is the geometry of the snow crystals. The water vapour pressure is greater over convex ice surfaces than over concave ones. Therefore, the tips of the crystals tend to sublimate (direct transition from the solid to the gaseous phase) and are degraded, the indentations are filled by desublimation (gaseous to solid). This water vapor transport results in increasingly spherical shapes, which also minimizes the surface-to-volume ratio and thus the surface energy.

Melt-freeze metamorphism naturally occurs only when temperatures fluctuate around 0 °C; it is the fastest and most efficient form of metamorphism. Wet snow is formed, in which the liquid water in the pores between the ice grains can be held against gravity. In wet snow, on the other hand, the pore space is so oversaturated with water that it runs off.

During the **constructive metamorphosis**, the particles enlarge and more complex cup-shaped crystals are formed, the so-called depth hoar. The prerequisite is a large temperature gradient of over 20 °C per metre in the snow cover, which leads to water vapour diffusion from warmer crystals to colder ones. The water vapour is deposited as ice on the colder crystal and begins to grow. Over time, this results in the formation of large faceted crystals, which are very cohesionless and often act as a sliding surface for slab avalanches. ◻ Figure 2.5 summarises all forms and products of snow metamorphosis.

2.2.3 Densification of the Snow

With the exception of the constructive metamorphosis, which only takes place when the snow cover is thin and temperatures are very cold, the air-filled pore space of the snow cover is reduced during snow metamorphosis. Thus, the density and hardness of the snow increase and the volume decreases, the snow cover "settles". The density of a substance is expressed as mass per volume in grams per

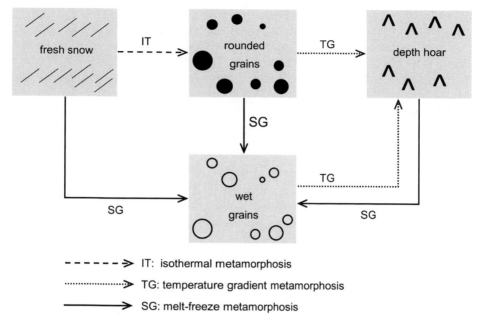

- - - - - - ➤ IT: isothermal metamorphosis

·················➤ TG: temperature gradient metamorphosis

─────────➤ SG: melt-freeze metamorphosis

▢ Fig. 2.5 Forms and products of snow metamorphosis

▢ Table 2.1 Hardness classes of snow (Fierz et al. 2009)

Designation	Hand test	Ram resistance (N)	Index
Very soft	Fist	0–50	1
Soft	Four fingers	50–175	2
Medium	One finger	175–390	3
Hard	Pencil	390–715	4
Very tough	Knife blade	715–1200	5
Ice	–	>1200	6

cubic centimetre (g cm^{-3}); liquid water has a density of about 1 g cm^{-3} (exactly: 0.999975 at 3.8 °C), which gives the weight of a litre of water as 1 kg.

The process of consolidation is also called **diagenesis,** by analogy with the formation of sedimentary rocks. Snow hardness is the resistance that snow offers to a penetrating object; it can be determined with penetrometers and expressed in Newtons (N). Alternatively, a five-grade index can be determined with a simple hand test. The decisive criterion here is whether the fist or correspondingly less can be pressed into a snow layer without great resistance (▢ Table 2.1).

2

Freshly fallen snow or **new snow** has a density of mostly less than 0.1 g cm^{-3} and consists of 90–97% air. Primary crystal structures are still visible, the consistency is often powdery ("powder snow"). Moist fresh snow, on the other hand, has density values between 0.1 and 0.2 g cm^{-3}.

Settled snow is snow that has already undergone metamorphic changes, which occur just a few days after snowfall. The air content then decreases significantly and the density increases to 0.2–0.4 g cm^{-3}.

One speaks of **firn** when the snow has lasted a whole summer. From this point on, further compression takes place only due to the overburden pressure. Firn has a broad density spectrum of 0.4–0.83 g cm^{-3}.

Glacial ice is present at a density of 0.83–0.917 g cm^{-3} (Paterson 1994). This can still contain up to 13% air, but unlike firn, the air bubbles are isolated and no longer form a coherent cavity. In other words, glacier ice is not permeable to air. Pure ice has a density of 0.917 g cm^{-3}, but in metamorphically formed ice there are always air inclusions and impurities, so that an average density of 0.9 g cm^{-3} can be assumed.

> With increasing age, new snow becomes settled snow and, if it has survived a summer, firn. Through further compression, the air content decreases and the density increases. As soon as it reaches a value of approx. 0.9 cm^{-3}, it is referred to as glacial ice.

How long it takes for glacier ice to form is strongly dependent on winter snowfall and summer temperatures. The less snow falls and the colder the summers, the longer the transformation takes because the overburden pressure increases more slowly and melt-freeze metamorphism occurs less strongly or not at all. From firn and ice cores it is known that with relatively high snow precipitation in the Central Alps (about 1500 mm per year) glacier ice can form after less than 20 years at a depth of less than 20 m (Oerter et al. 1982). At Camp Century in Greenland (annual snow accumulation: 320 mm), glacier ice only forms after 125 years at a depth of 68 m (Gow 1971), and at Dom C in Antarctica, which has very low precipitation (annual snow accumulation: 36 mm), it takes a full 1700 years for compact glacier ice to form at a depth of 100 m (Raynaud et al. 1979, cited in: Paterson 1994). At year-round cold temperatures, no melt-freeze metamorphosis occurs here and, in addition, water vapour diffusion is slowed down.

> The duration of formation of glacial ice can range from a few years to over 1000 years. It lasts longest when no melting occurs and very little snow falls. In the Alps, metamorphosis is usually completed in less than 20 years.

2.3 Physical Properties of Glacial Ice

Knowledge of the physical properties of this medium is indispensable for the understanding of many glaciological phenomena or for questions dealing, for example, with the movement of ice. For this purpose, even the smallest structural

☐ Fig. 2.6 Structure of an ice crystal in plan view (**a**) and in elevation (**b**). (From Paterson 1994; courtesy © Elsevier AG 1994, all rights reserved)

units, namely molecules, must be considered, because their structure has direct consequences for the deformation of ice (▶ Chap. 3).

Ice crystallizes according to the molecular structure of water in hexagonal form, with the oxygen atoms are arranged in regular hexagons and on two planes (unfilled and filled circles in ☐ Fig. 2.6a), which are 0.923 Å (Ångström; 1 Å corresponds to the ten-millionth part of a millimetre) away from each other (☐ Fig. 2.6b).

The next layer, in which the atoms are arranged in mirror image, is located at a distance of 2.760 Å. This parallel-layered structure results in a plate-like crystal structure due to the arrangement in basal planes. Glacial ice does not react in the same way in all directions to physical stresses or the incidence of light. For example, it is only completely transparent perpendicular to the basal planes; in all other directions, the light is refracted differently. Also, deformation due to stress is not the same in all directions, which will play a role in ▶ Chap. 3. Such a direction-dependent behaviour of a material is called **anisotropy.**

❯ The water molecules are arranged in planes in an ice crystal, resulting in a platy crystal structure.

Glacial ice has a granular structure, whereby grains can consist of single or multiple crystals. The size of the ice crystals increases with time; Forel (1882) already put this growth at 1.4% per year. As the age of the ice increases downwards, the largest crystals are found at the glacier tongue. Diameters of 8 cm have been observed in

2

◘ **Fig. 2.7**   Large single crystals in immobile ice above a bergschrund. (Photo: W. Hagg)

◘ **Fig. 2.8**   Layering of ice in a glacier cave at Vernagtferner. (Photo: W. Hagg)

the Alps, maximum values of 1.5 cm are reported from polar regions (Charlesworth 1957). Strong ice movement crushes ice crystals, so that the largest specimens are found in the calmer, marginal areas. Single-grain diameters of 18 cm have been recorded in unmoving **dead ice** (Seligman 1948). Very large crystals can also form above the **bergschrund,** where the ice is too thin to move (▶ Sect. 3.4) (◘ Fig. 2.7).

The texture of glacial ice often shows stratification caused by snowfall events, dust inputs and the difference between summer and winter snow. Summer snow is turned into air-poor, bluish ice by melting events, whereas winter snow is compacted into air-rich, white ice (◘ Fig. 2.8). In the upper parts of the glacier, where the ice is formed, these layers lie parallel to the surface; the movement of the ice tilts them towards the end of the glacier (▶ Chap. 3).

In addition to stratification, **foliation** can also occur, especially in polar regions. This also results in an alternating banding of air-rich, coarse crystalline ice and air-poor, fine crystalline ice . These are pressure textures produced by an orientation of the platy crystals perpendicular to the stress. They can have a relationship to stratification and are often difficult to distinguish from it.

The hardness of ice is temperature-dependent and reaches its maximum at −70 °C with a mineralogical hardness of six, which corresponds to that of feldspar.

❯ The melting point of glacier ice is pressure-dependent. At atmospheric pressure it is 0 °C, within the ice it decreases by 0.0073 °C per bar increase in pressure and is referred to as the **pressure melting point.**

For an ice thickness of 120 m, the pressure melting point is about −0.1 °C; for mountain glaciers, it always remains above −1 °C. Only at an ice thickness of 2000 m, as can occur in the polar ice sheets, is the load-induced pressure melting point −1.6 °C (Bennet and Glasser 2009).

Melting point and temperature thus change from the glacier surface to the base, where the glacier rests on the rock. At the surface, the ice temperature is largely controlled by the air temperature. In summer it is 0 °C under melting conditions, and in winter it can cool well below this value. At the base of the glacier, the ice temperature results from the balance of heat gain and loss.

Heat gain results from the geothermal heat flux from the Earth's interior and from frictional heat, both within the ice and between the ice and the subsurface. On average, the frictional heat is equivalent to the mean geothermal heat flux for an ice movement of 20 m per year (Bennet and Glasser 2009).

Heat loss is caused by the removal of energy towards the surface. This energy transport is controlled by the temperature gradient (°C per metre) between the base and the surface and by the thermal conductivity of the glacier ice. The gradient depends on the surface temperature and the thickness of the ice. The colder the air temperature and thus the temperature of the ice surface and the thinner a glacier is, the greater the gradient and the more effectively heat can be dissipated from the base. As a result, the base cools down, so that there is a direct correlation between the temperature at the surface and at the glacier bed: a cooling or warming of 1 °C at the ice surface causes, with a delay, the same temperature change at the base.

If the energy gain at the base is greater than the removal, net melting occurs; in the opposite case, net freezing occurs. Two basic vertical temperature profiles result, which are shown in ◨ Fig. 2.9.

In the case of **cold ice,** as occurs mainly in polar regions, the temperature at any depth is well below the pressure melting point (◨ Fig. 2.9a). The temperature decreases continuously from the base of the glacier to the surface, allowing efficient energy dissipation at the base where, consequently, meltwater does not occur. In **temperate ice,** which is the normal case in mid-latitude glaciers, the temperature is at the pressure melting point over the entire profile (◨ Fig. 2.9b) and meltwater occurs at the base of the glacier. These temperature conditions at the glacier base have decisive effects on ice movement (▶ Chap. 3) and glacial erosion (▶ Chap. 9).

2

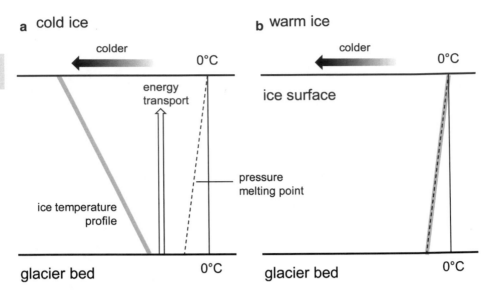

a cold ice   **b** warm ice

◘ Fig. 2.9   Idealized temperature profile through cold ice (**a**) and warm ice (**b**). (Modified after Chorley et al. 1984)

References

Bennet MR, Glasser NF (2009) Glacial geology. Ice sheets and landforms, Wiley-Blackwell, Chichester

Charlesworth JK (1957) The quaternary era, Bd 1. Edward Arnold, London

Chorley RJ, Schumm SA, Sugden DE (1984) Geomorphology. Methuen, London/New York

de Martonne E (1948) Traité de Géographie physique. Armand Colin, Paris

de Quervain M (1963) On the metamorphism of snow. In: Kingery WD (ed) Ice and snow. MIT Press, Cambridge, MA, pp S 377–S 390

Fierz C, Armstrong R, Durand Y, Etchevers P, Greene E, McClung D, Nishimura K, Satyawali P, Sokratov S (2009) The international classification for seasonal snow on the ground. Technical documents in hydrology, vol 3. UNESCO, Paris

Forel FA (1882) Le grain du glacier. Arch Sci 3(7):329–375

Furukawa Y (1997) Faszination der Schneekristalle – wie ihre bezaubernden Formen entstehen. Chemie in unserer Zeit 31(2):58–65

Gow AJ (1971) Depth-time-temperature relationships of ice crystal growth in polar glaciers. CRREL Res Rep 300:1–19

Nakaya U (1954) Snow crystals, natural and artificial. Harvard University Press, Cambridge, p S 510

Oerter H, Reinwarth O, Rufli H (1982) Core drilling through a temperate alpine glacier (Vernagtferner, Oetztal Alps) in 1979. Z Gletscherk Glazialgeol 18(1):1–11

Paterson WSB (1994) The physics of glaciers. Butterworth Heinemann, Oxford/Burlington, p S 481

Raynaud D, Duval P, Lebel B, Lorius C (1979) Crystal size and total gas content of ice: two indicators of the climatic evolution of polar ice sheets. In: Evolution of planetary atmospheres and climatology of the earth, Proceedings of the international conference held 16–20 October, 1978 in Nice, France. Centre National d'Etudes Spatiales, Toulouse

Seligman G (1948) The growth of the glacier crystal. IAHS Publ 29:216–220

Weischet W (1991) Einführung in die Allgemeine Klimatologie. Teubner, Stuttgart

Ice Movement

Contents

3

Overview
While the general movement patterns of glaciers were already described at the end of the nineteenth century, the three sub-processes that enable the ice to move were only recognised later. These are internal deformation, basal sliding and bed deformation. Glacial surges, although not a form of movement in their own right, are nevertheless a special case of ice movement because of their speed. The different types of glacial surges occur in places where flow velocities vary; they allow conclusions to be drawn about ice movement and bedrock. Together with the ogives, crevasses are visible and conspicuous evidence that glacial ice is not rigid, but to a certain extent deformable in response to stress.

3.1 Description of the Movement Pattern

That glacial ice is not a static mass but can move has been known to man at least since the glacial advances of the Little Ice Age (▶ Chap. 8). But even stationary glaciers, which are in equilibrium with the local climate and whose fronts remains in the same place for several years, are in constant motion. In the altitudinal regions where glacier ice is formed by metamorphism, there is a mass surplus because more snow accumulates there than melts. This mass surplus is transported downslope by ice movement, where it compensates for the mass deficit caused there by the preponderance of melt (▶ Chap. 4). The first measurements of movement and attempts to explain this phenomenon were made as early as the eighteenth century (▶ Chap. 1). The first to describe ice motion in detail and correctly was Sebastian Finsterwalder at Vernagtferner in Austria. He recognized that for every point above the glacier snowline where a snowflake falls, there corresponds a point below the snowline where it melts again (◘ Fig. 3.1). This nice idea does not take into account snow metamorphosis and the fact that snow can also melt above the equilibrium line, but the conclusions on the general trajectories are quite correct.

The lines of motion run from top to bottom along the solid lines in ◘ Fig. 3.1. The equilibrium line is indicated by the dashed line. A snowflake that falls here melts again at the same point. The lines of motion converge, i.e. narrow, to the equilibrium line, and they widen (diverge) below it. The entire firn field, which is the area above the equilibrium line, is mirrored in the ablation area. Finsterwalder further recognized that the two aforementioned points of snow deposition and ice melt are connected by a line running inside the glacier, which he calls the streamline (◘ Fig. 3.2).

On the dark blue marked area above the firn line, mass gain takes place, on the red area below mass loss predominates. The dashed lines are streamlines and show the movement paths within the glacier. Above the equilibrium line, they run downwards, i.e. into the interior of the glacier (blue arrows). This inward movement is called **submergence**. The angle to the surface increases towards the top because the annual snowfall increases with increasing sea level. The solid lines are layers that represent former surfaces of the firn area. They are deformed by the

□ Fig. 3.1 Top view of a glacier. The solid lines show the ice movement, the dashed lines are layer lines of equal mass change. Snowflakes falling at the marked points disappear into the glacier and reappear at the points marked with drops. (After Finsterwalder 1897, supplemented)

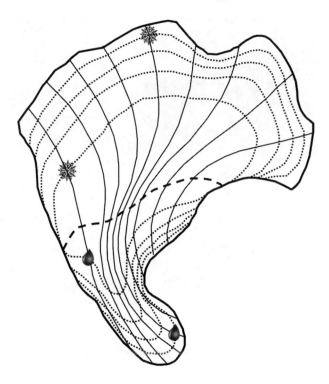

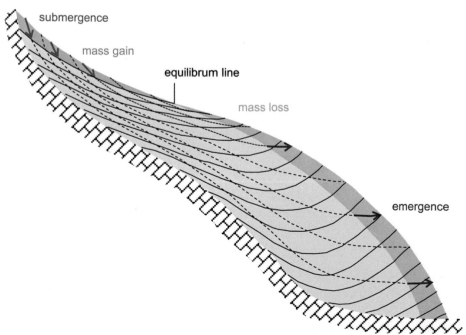

□ Fig. 3.2 Zones of mass gain and mass loss as well as movement vectors and layer boundaries in the longitutinal section of a glacier. (After Finsterwalder 1897, supplemented)

3

◘ Fig. 3.3 Remains of an aircraft on the Watzmann Glacier (left) and a World War II grenade at Nördlicher Schneeferner (right). (Photos: W. Hagg)

glacier as it moves. Below the firn line, the ever-increasing melt results in a relative movement against the surface called **emergence** (red arrows). The streamlines run towards the ice surface, and the angle of emergence of the streamlines increases downwards. The surface layers are also straightened up towards the end of the glacier. Finsterwalder (1897) already postulated in his "geometric-kinematic theory of motion" that the exit angle of the surface layers corresponds to the sum of the angle of dip and the exit angle of the streamlines. In simplified terms, he recognized the following: The further above the firn line a particle falls onto the glacier, the deeper it is transported into the ice, and the further down it reaches the surface again. It does not necessarily have to be a snowflake or a stone; lost objects, dead mountaineers or crashed aircraft are also transported through the ice along these lines (► Excursus 3.1) (◘ Fig. 3.3).

> Everything that falls on the glacier above the equilibrium line disappears into it. The higher up this happens, the longer it disappears and the further down it comes back to the surface. Everything that lies on the glacier below the equilibrium line remains on the surface until it reaches the front of the glacier.

In the top view (◘ Fig. 3.1) it can also be seen that the outcrops of the former surfaces (dashed) below the firn line are bent downwards in the centre of the glacier. These so-called "**false ogives**" (the "true ogives" are discussed in ► Sect. 3.2.4) show that ice motion is greater there than at the margins, where frictional influence increases. An example of false ogives can be seen in ◘ Fig. 3.4. The Höllental-ferner flows from the right, where a snow and firn layer can still be seen, to the

◘ **Fig. 3.4** Transverse crevasses and "false ogives" on Höllental-ferner. (Photo: W. Hagg)

left. The surfaces of the individual years can be seen as different shades of grey in the right part of the image. To the left of the crevasse zone, these are then steeply tilted-up and can only be seen as linear structures (outcrops) on the surface. These are bent downwards in the middle of the glacier because the ice flows faster here than at the edge.

At Höllentalferner, these false ogives show nicely that ice can deform plastically. At the same time, the neighbouring crevasses show that it is not deformable at will, but can also react brittly to stress and crack. Sebastian Finsterwalder was able to describe the movement of glaciers, but could not yet explain it. Some time should pass before the development of modern theories of ice movement, which are described in the next section.

Excursus 3.1: Finds in the Ice

Nothing disappears forever in the glacier, but for some things it takes quite a while until they reappear. In this context, the finds can allow conclusions to be drawn about the flow rate of the ice or, conversely, knowledge of the ice movement can allow the transport to be reconstructed. Jouvet and Funk (2014) succeeded in calculating the location of the climber's accident in 1926 from the discovery of two identified corpses on the Aletsch Glacier in 2012 and even deduced the cause of the accident.

Often the finds are remnants from wartime, such as ammunition, shells or aircraft parts. Since 2003, the remains of a Junkers JU-52, which crashed on the Watzmann east face in October 1940, have been melting out on the Watzmann glacier (◘ Fig. 3.3).

3

The most financially spectacular discovery are probably gems said to be worth several hundred thousand dollars, found in 2013 on the Bossons glacier near Chamonix. These come from the crash of an Air India plane in 1966 (BBC News 2013). The mention of the year is important here in that two Air India planes had already crashed on the same glacier.

A large number of finds, although not as valuable in monetary terms, are significantly older and gave rise to the relatively young special field of glacial archaeology, which already has its own journal. Roman coins, tools from the Iron Age and weapons from the Early Bronze Age have been found on the Lötschen Glacier in the Bernese Alps (Hafner 2015).

But glacial corpses can also be of astonishing age. A mercenary who died 400 years ago was discovered on the Theodul Pass (Providoli et al. 2015), and in the Canadian Rocky Mountains an American Indian, complete with cape and hat, who had died 550 years earlier, turned up in 1999 (Beattie et al. 2000). The most famous find, however, was certainly the Ötztal Iceman. However, contrary to popular belief, this mummy was not released from a glacier, but melted out of a perennial firn patch. This was a stroke of luck, as moving glacier ice would certainly have taken a much greater physical toll on Ötzi.

It remains to be seen what secrets the glaciers will reveal in the coming years; an absolute highlight for glacial archaeologists would probably be an elephant from Hannibal's caravan in 218 BC.

3.2 Processes Involved in Ice Movement

Today we know that three processes can be responsible for ice movement. While the first process occurs at every glacier, the other two are linked to the presence of certain conditions.

3.2.1 Internal Deformation

Ice is subject to **shear stress**, which depends on the pressure of the overlying ice and the surface slope of the glacier. This shear stress acts both on individual ice crystals and on a body of several crystals.

The deformation of single crystals was investigated in the laboratory. It was found that deformation takes place along the basal planes in the ice crystal (▶ Sect. 2.3) (Paterson 1994). Sliding processes can easily take place along these planes, as in a stack of new playing cards, so that even small shear stresses cause deformation.

In polycrystalline ice, the orientation of the basal planes in the individual crystals is random, which is why they mutually hinder deformation and make movement difficult. However, if the shear stress reaches a value of 50 kPa (kilopascal), then the **internal deformation** of polycrystalline ice begins through various partial processes in the crystal association. In addition to sliding processes within the ice crystals, these include rotational movements, directional growth of crystals and minute cracks that lead to misalignment. The complete recrystallization of crystals, whose basal planes are then optimally oriented for gliding processes, also plays a major role.

As a polycrystalline, anisotropic body, glacier ice reacts in a complex manner to pressure or shear stress. It is neither viscous nor plastic. If viscous, the deformation would increase linearly with the load (◨ Fig. 3.5); if plastic, there would be no deformation below a threshold (yield stress), above which it would correspond to a viscous fluid. The behaviour of glacial ice can best be described by Glen's flow law (1955), according to which deformation increases exponentially with increasing stress:

$$\varepsilon = A \ \tau^{n} \tag{3.1}$$

- ε = Deformation rate
- A = Temperature-dependent constant
- τ = Shear stress
- n = Exponent, depending on crystal orientation, crystal size, etc.

Glen's flow law is not a physical law, but an empirical equation derived by experiment. It fails at small stresses, but above about 50 kPa it is still the best basis for modelling ice motion. The constant A increases with increasing temperature, and the shear stress t even has an exponential effect on the deformation rate. Thus it follows from Glen's law: Glacial ice becomes more fluid with increasing stress and increasing temperature.

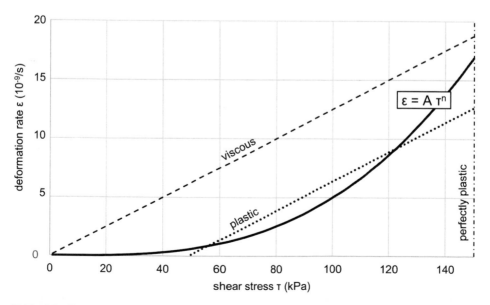

◨ **Fig. 3.5** Representation of different flow behaviours. Dashed line = linear viscous flow of a Newtonian fluid (viscosity: 8×10^{9} kPa), dotted line = plastic material with a yield stress of 50 kPA (viscosity: 8×10^{9} kPa), dashed dotted line = perfectly plastic material with a yield stress of 150 kPa (viscosity: against ∞), solid line = deformation of ice according to Glen's law (n − 3, A = 5×10^{-15})

3

> Glacial ice does not deform until a certain stress is reached, but then the deformation increases exponentially to the shear stress. Moreover, the closer its temperature is to 0 °C, the more deformable ice is.

If we consider the pressure-dependent deformation in the glacier profile (◘ Fig. 3.6a), then deformation (dotted grey line) only begins at a certain depth, then increases exponentially and goes back to zero directly above the bedrock due to the frictional drag. If the deformation and thus the flow velocity are considered individually for each depth, the maximum is thus located relatively close above the glacier bed. However, since the internal deformation at a certain depth also moves the overlying ice, the total motion (grey line in ◘ Fig. 3.6b) at each depth is always the sum of all motion vectors from the glacier bed to the depth considered. The motion vectors are to be considered cumulative from bottom to top, therefore the largest total motion always takes place at the surface. Although no deformation takes place here due to the lack of stress, the ice at the surface is transported by the movement at each depth (◘ Fig. 3.6b).

Internal deformation is the most natural form of movement of glacial ice, it occurs on all glaciers. It does not even require steep relief, even if the ground is perfectly flat, an ice body starts to flow laterally from a certain thickness (the one that generates at least 50 kPa stress inside it). Internal deformation creates a lami-

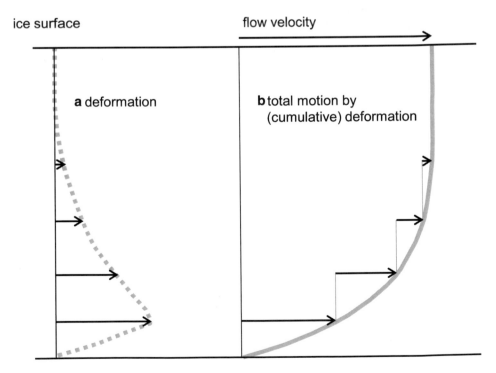

ice surface   flow velocity

a deformation   **b** total motion by
                                    (cumulative) deformation

glacier bed

◘ **Fig. 3.6** Deformation (**a**) and total motion (**b**) in longitudinal section of a glacier

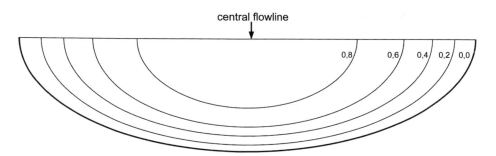

Fig. 3.7 Theoretical velocity profile due to deformation in an elliptical glacier cross-section (width to depth ratio 4:1). The figures are dimensionless velocity measures. (After Nye 1965)

nar flow pattern that can be compared to a slow-moving river. "Laminar" means that the lines of motion do not cross but are parallel. Due to the frictional drag, flow velocities are lowest at the margin; glaciers flow fastest where they are most powerful. In the case of mountain glaciers, this is usually the middle of the valley (■ Fig. 3.7); at constrictions – as in rivers – particularly high velocities are reached.

3.2.2 Basal Sliding

The second form of movement, basal sliding (Weertmann 1964), refers to the sliding of the ice across the glacier bed. In this mode of motion, the glacier slides in one piece (en bloc), so that the velocity is the same over the entire vertical profile.

Basal sliding can only occur when there is a film of meltwater between the ice and the bedrock. This significantly reduces the number of contact points and thus the friction between the two materials and even allows the ice to float slightly for a short time. Because basal melting is necessary for the type of movement, it can only occur with temperate or warm ice (▶ Sect. 2.3), which is at the pressure melting point. Rock obstructions are overcome by plastic flow around them (enhanced basal creep) or by regelation slip. The former is based on the fact that the plasticity and thus the ability to flow around the obstacle is promoted by an increase in pressure on the upper side. In the case of regelation slip, increased pressure melting occurs on the upglacier side (stoss), whereby the water passes the rock bar in liquid form and freezes again on the downglacier side (lee) due to the decrease in pressure.

Basal sliding processes are seasonal and have their maximum when there is a particularly high occurrence of meltwater, i.e. in summer and especially during hot spells. Some of the supraglacial water finds its way under the glacier and increases the basal water pressure there. Especially high pressures in subglacial cavities reduce the roughness of the bed, facilitating sliding processes.

In polar glaciers that are frozen to the bedrock, this form of movement is not completely absent, but it is so small that it can be neglected (Fowler 1986).

> In basal sliding, the glacier as a whole slides over its bedrock. This requires a film of meltwater, which is why this movement type only occurs in temperate ice.

3.2.3 Bed Deformation

Glacier beds consisting of water-saturated unconsolidated sediment can be deformed by the pressure of the ice, which in turn moves the overlying glacier ice. Liquid water exists in subglacial deposits only under warm ice, which is why this type of movement, like basal sliding, occurs only in temperate glaciers. The movement, which is also en bloc in this case, is determined less by the properties of the ice than by mechanical and hydrological properties of the subsurface. Deformation of the sediment can increase or decrease downslope; this also appears to be partly controlled by water saturation at different depths in the loose substrate (Evans et al. 2006). At Black Rapids Glacier in Alaska, no deformation occurs at all in the uppermost two sediment layers, but it must be substantial in deeper layers, where half to three-quarters of the total movement is known to be caused by sediment deformation (Truffer et al. 2000). Many relationships are still unexplored, because the processes involved are complex and naturally very difficult to observe. In extreme cases, as at Breiðamerkurjökull in Iceland, bed deformation can account for 90% of the total movement (Boulton and Hindmarsh 1987).

> When glaciers lie on water-saturated loose material, this can be deformed by the pressure of the ice, causing the glacier to move with it. The process is difficult to observe and quantify.

3.2.4 Glacier Flow Velocity

The flow velocity of mountain glaciers can vary greatly between 1 and 800 m per year (Boulton 1974). The highest values are reached at narrows or at steep steps, and very low values at glacier tongues that are heavily covered with debris. Just as movement varies spatially, it does so temporally. Most glaciers have seasonal variations with higher velocities in the melt period, indicating the importance of basal sliding. At the Southern Inylchek glacier in the central Tienshan, the surface flow velocity 14 km upstream of the glacier terminus is 1.5–2 times higher in summer than in the annual mean, where it is 130 m per year in the main flowline and 80–90 m per year averaged across the cross-section (Mayer et al. 2008). At the same glacier, the movement at the heavily debris-covered glacier end approaches zero.

At the Argentière glacier in France, the movement of the glacier is measured in real time in a cavity beneath the glacier. Here it could be shown that ice movement increases after heavy precipitation, which is probably due to the increased water pressure between ice and rock and thus an increase in basal sliding (Benoit et al. 2015). Mean flow velocity reaches about 70 m per year at the relatively steep monitoring site, which is half of what it was in the 1980s (Vincent and Moreau 2016). At other locations, too, mountain glaciers are tending to slow down in the course of the current glacier retreat, because the ice supply from the accumulation area is decreasing.

◘ Fig. 3.8 The Perito Moreno glacier flows into Lago Argentino, reaching velocities of several hundred metres per year. (Photo: Clemens Netzer, January 2020)

Comparatively high flow velocities are found on glaciers that flow into lakes or the sea. Due to the upwelling of the glacier tongues, values of 620 m per year are reached, for example, at the Perito Moreno Glacier in Argentina (◘ Fig. 3.8) (Minowa et al. 2017). The Jakobshavn Isbræ in Greenland, which also calves into the sea, is considered the fastest glacier in the world, with ice moving at peak speeds of up to 17 km per year after an extreme melt period in the summer of 2012 (Joughin et al. 2014), but has since (2015–2017) slowed back down to 1250 m per year (Lemos et al. 2018).

Cold glaciers, in contrast, move very slowly. On the one hand, there is practically no basal sliding, on the other hand, the temperature-dependent internal deformation in polar regions is also strongly slowed down.

> ❯ Glaciers have a wide range of speeds, from a few metres to several hundred metres per year; glaciers that flow into the sea can be even faster. In the course of glacier retreat, the flow velocity is currently mostly decreasing.

3.3 Special Case Surge

In some glaciers, phases of strongly accelerated ice movement occur. At the aforementioned Vernagtferner in the Ötztal Alps, four very rapid advances occurred during the Little Ice Age between 1450 and 1850 AD (▶ Chap. 8), during which the glacier advanced into the main valley at speeds of up to 11.5 m per day (Nicolussi 2013). There, the glacier formed a dam and gave rise to the Rofen ice-dammed lake(◘ Fig. 3.9).

The lake partially emptied due to dam breaches, which resulted in devastating flood disasters in the Ötztal. For this reason, the Vernagtferner was already early in the awareness of the public and science, possibly also for this reason Sebastian Finsterwalder chose it for his mapping at the end of the nineteenth century (▶ Chap. 1). Glaciers with such advance phenomena used to be called "galloping" or "pulsating" glaciers; today they are referred to as glacial surges.

3

□ Fig. 3.9 The Vernagtferner, which has disintegrated into blocks, dams up the Rofen ice-dammed lake in 1771. (Colour lithograph by C.F. Hoppe, printed in 1810 by Matthies, Schmiedberg)

During these periodic advances, the glaciers flow ten to 1000 times faster than in the phases in between and reach maximum velocities of several tens of metres per day. The quiescent phases typically last 15–100 years, the advance phase only 1–10 years (Meier and Post 1969). The phenomenon is observed in less than 1% of all glaciers; it occurs in some regions (Alaska, Canada, Greenland, Iceland, Spitsbergen, Pamir, Karakoram), in others not at all. The cumulation in subpolar regions and in very high mountains, where the formation of polythermal glaciers (▶ Chap. 5) is favoured, is striking. Here, the upper parts often consist of warm ice, while the lower parts of the glacier tongue may be cold. This impedes englacial and subglacial runoff, which can lead to water accumulation and large-scale upwelling of the glacier.

Another theory assumes an obstruction of the ice flow, for example by a rock obstacle. Above this partial blockage, there is an increase in ice thickness. This also increases the pressure on the subglacial channel system (▶ Chap. 7) and the water is forced out of the central drainage channel into the contact zone between rock and ice. At a certain point, the subglacial drainage system switches from a linear system to an areal drainage system (linked cavity theory after Kamb 1987). The ice floats up, friction is strongly reduced and basal sliding increases abruptly. During the surge of Variegated Glacier in Alaska, this form of motion accounted for about 90% of the total motion according to Kamb et al. (1985). As a result, the pent-up

◘ Fig. 3.10 The bulging and highly fractured front of Shisper Glacier in Pakistan during a surge in 2019, having advanced 1 km in half a year. (Photo: Astrid Lambrecht)

mass imbalance can be relocated relatively suddenly. This occurs in the form of a thickening that undulates throughout the glacier, causing a sudden advance once it reaches the glacier terminus. Due to the large increase in basal sliding and the accompanying en bloc motion, the glacier breaks into individual blocks. This phenomenon is particularly well visible in ◘ Fig. 3.8.

A third theory suggests that a surge is supported by the strong deformation of soft ground (Boulton 1979).

Recent examples of surge events can be found in High Asia. At Bivachny Glacier in the Pamirs, an 80 m thick thickening moved 13 km through the glacier tongue at a rate of 4400 m per year from 2011 to 2015 (Wendt et al. 2017). Similarly, at Hispar Glacier in the Karakoram, surging sections have moved at speeds of up to 900 m per year during 2013–2017. In May 2017, the Khurdopin Glacier in the Karakoram advanced at peak velocities of over 5000 m per year after an 18-year dormant period, damming a lake (Steiner et al. 2018). The Shisper Glacier in the Karakoram makes a particularly threatening appearance during a current surge (◘ Fig. 3.10).

❯ Glacial surges are only a mechanical redistribution and not an increase in mass. Since such advances have no climatic cause, surging glaciers are unsuitable as climate indicators.

3.4 Visible Witnesses of Ice Movement: Crevasses and Ogives

Ice is not deformable at will. Above a certain shear stress, the deformation can no longer take place fast enough and the ice breaks. This is how crevasses are formed. As already shown in ▶ Sect. 3.2.1, deformability depends on temperature and pressure according to Glen's flow law. The stress and thus also the plasticity increase with depth. From a certain depth, ice is so rapidly deformable that crevasses no longer form. There is thus a maximum depth of crevasses that depends on the second factor for deformability, temperature. In the mid-latitudes, this value is about 30 m; the cold and therefore more brittle ice of the polar regions can break up to depths of about 90 m. Crevasses occur preferentially at certain spots on the glacier and can be classified according to these locations.

The highest crevasse is the **Bergschrund**, which marks the boundary between frozen and moving ice. In the uppermost glacier areas, the ice is often still too thin to be moved by internal deformation. When such areas are composed of cold ice and have no meltwater film at the base, they are frozen to the bedrock and immobile. Above a certain thickness, the stress is sufficient for deformation and the ice begins to flow, detaching from the immobile part in the form of a crevasse (◘ Fig. 3.11).

A very common type of crevasse is the **transverse crevasse**. It occurs at valley steps where the slope and thus the flow velocity of the ice suddenly increase. Such steep steps are also called **ice falls.** At the upper, convex bend, the glacier is stretched, which is called **extending flow.** If the differences in velocity are so great that the ice cannot deform fast enough, it cracks and transverse crevasses form (◘ Fig. 3.12). When the crevasses become very wide, the ice between them can break into individual towers called **séracs.** At the concave bends, where the slope decreases, faster ice flows onto slower ice; in such places with **compressive flow**, transverse crevasses do not form (◘ Fig. 3.12).

Marginal crevasses occur at lateral glacier boundaries. Here, a tension is built up due to the decrease in velocity towards the edge. This can be easily reproduced with the help of a tablecloth: If one pushes the cloth forward more with one hand (center of the glacier) than with the other hand (edge of the glacier), folds are created that are oriented at 45° to the center of the glacier in the direction of motion.

◘ **Fig. 3.11**   Bergschrund in the Caucasus. (Photo: W. Hagg)

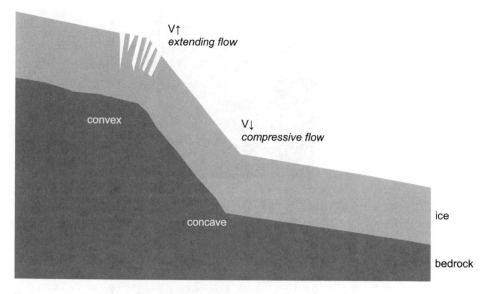

Fig. 3.12 Schematic of the formation of transverse crevasses on terrain irregularities

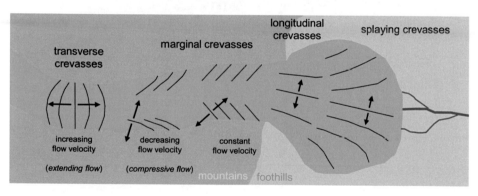

Fig. 3.13 Typical crevasse patterns along a glacier. (Modified after Nye 1952 and Hambrey and Alean 2004)

Strain stresses act transversely to these folds, causing crevasses to tear open. These crecasses are therefore perpendicular to the creases in the tablecloth experiment and point with 45° against the direction of movement. These theoretical geometric considerations apply to homogeneous conditions with respect to glacier bed and flow velocity; in the case of a slowdown (compressive flow), the angle of the crevasses becomes flatter towards the centre of the glacier (Fig. 3.13).

Longitudinal crevasses occur when expansion stresses occur transversely to the direction of flow. This can occur when the valley suddenly widens. With less lateral constriction by the valley flanks, longitudinal crevasses then tear open if the strain is faster than the maximum deformation rate of the ice. When a glacier leaves the mountains, it flows out in a lobate manner and forms **splaying crevasses** in all directions according to the elongation (Fig. 3.13).

▪ Fig. 3.14 Ogives below an escarpment on Gates Glacier, Alaska. (Google Earth)

> Crevasses always occur where differences in flow velocity are greater than the deformation capacity of the ice. Wide crevasses are always caused by stretching movements, while small crevasses can also be caused by shear movements.

In zones with strongly decreasing flow velocities, for example below ice falls, compressive bulges can form transverse to the flow direction. Due to the higher flow velocity in the middle of the glacier, these surface structures are usually curved and then called **ogives** after a Gothic arch shape (▪ Fig. 3.14).

Such bulging wave ogives are distinguished from band ogives, in which lighter and darker bands alternate without the surface becoming wavy. Both types can also overlap, further complicating the phenomenon. In addition, wave-like topography can develop secondarily due to varying degrees of melting on the lighter and darker bands. Both types of ogives are seasonal formations and are due to the difference in flow velocity between summer and winter. A bulge may be caused by faster flow in summer and the resulting increased compression at the base of the escarpment. In the case of color banding, an ogive consists of a light and a dark band, and there are several theories as to its origin. Both seasonally varying debris or dust concentrations and uplifting of dark, basal ice through shear zones (Goodsell et al. 2002) have been suggested as causes.

References

BBC News (2013) Alpine climber finds, India plane crash' jewels. https://www.bbc.com/news/world-europe-24294330. Accessed on 14.04.2020

Beattie O, Apland B, Blake EW, Cosgrove JA, Gaunt S et al (2000) The Kwädäy Dan Tsìnchi discovery from a glacier in British Columbia. Can J Archaeol 24:129–147

Benoit L, Dehecq A, Pham H, Vernier F, Trouvé E, Moreau L, Martin O, Thom C, Pierrot-Deseilligny M, Briole P (2015) Multi-method monitoring of Glacier d'Argentière dynamics. Ann Glaciol 56(70):118–128. https://doi.org/10.3189/2015AoG70A985

Boulton GS (1974) Processes and patterns of glacial erosion. In: Coates DR (ed) Glacial geomorphology. State University of New York, Binghamton, pp S 41–S 87

Boulton GS (1979) Processes of glacier erosion on different substrata. J Glaciol 23(89):15–38

Boulton GS, Hindmarsh RCA (1987) Sediment deformation beneath glaciers: rheology and geological consequences. J Geophys Res 92:9059–9082

Evans DJA, Phillips ER, Hiemstra JF, Auton CA (2006) Subglacial till: formation, sedimentary characteristics and classification. Earth Sci Rev 78(1–2):115–176. https://doi.org/10.1016/j.earscirev.2006.04.001

Finsterwalder S (1897) Der Vernagtferner. Wiss Ergänzungshefte Z Dtsch Osterr Alpenvereins 1(1): 1–112

Fowler AC (1986) Sub-temperate basal sliding. J Glaciol 32(110):3–5. https://doi.org/10.3198/1986JoG32-110-3-5

Glen JW (1955) The creep of polycrystalline ice. Proc R Soc A 228(1175):519–538

Goodsell et al (2002) Formation of band ogives and associated structures at Bas Glacier d'Arolla. Valais, Switzerland. https://doi.org/10.3189/172756502781831494

Hafner A (2015) Schnidejoch und Lötschenpass. Archäologische Forschungen in den Berner Alpen. Archäologischer Dienst des Kantons Bern, Bern

Hambrey M, Alean J (2004) Glaciers. Cambridge University Press, Cambridge

Joughin I, Smith BE, Shean DE, Floricioiu D (2014) Brief communication: further summer speedup of Jakobshavn Isbræ. Cryosphere 8:209–214. https://doi.org/10.5194/tc-8-209-2014

Jouvet G, Funk M (2014) Modelling the trajectory of the corpses of mountaineers who disappeared in 1926 on Aletschgletscher, Switzerland. J Glaciol 60(220):255–261. https://doi.org/10.3189/2014JoG13J156

Kamb B (1987) Glacier surge mechanism based on linked cavity configuration of the basal water conduit system. J Geophys Res 92(B9):9083–9100

Kamb B, Raymond CF, Harrison WD, Engelhardt H, Echelmeyer KA, Humphrey N, Brugman MM, Pfeffer T (1985) Glacier surge mechanism: 1982–1983 surge of Variegated Glacier, Alaska. Science 227:469–479

Lemos A, Shepherd A, McMillan M, Hogg AE, Hatton E, Joughin I (2018) Ice velocity of Jakobshavn Isbræ, Petermann Glacier, Nioghalvfjerdsfjorden and Zachariæ Isstrøm, 2015–2017, from Sentinel 1-a/b SAR imagery. Cryosphere 12:2087–2097. https://doi.org/10.5194/tc-12-2087-2018

Mayer C, Lambrecht A, Hagg W, Helm A, Scharrer K (2008) Post-drainage ice dam response at Lake Merzbacher, Inylchek glacier, Kyrgyzstan. Geogr Ann 90 A(1):87–96

Meier MF, Post A (1969) What are glacier surges? Can J Earth Sci 6:807–817

Minowa M, Sugiyama S, Sakakibara D, Skvarca P (2017) Seasonal variations in ice-front position controlled by frontal ablation at Glaciar Perito Moreno, the Southern Patagonia Icefield. Front Earth Sci 5:1. https://doi.org/10.3389/feart.2017.00001

Nicolussi K (2013) Die historischen Vorstöße und Hochstände des Vernagtferners 1600–1850 AD. Z Glaziol Glazialgeol 45(46):9–23

Nye JF (1952) The mechanics of glacier flow. J Glaciol 2:82–93. https://doi.org/10.3198/1952JoG2-12-82-93

Nye JF (1965) The flow of a glacier in a channel of rectangular, elliptic, or parabolic cross-section. J Glaciol 5(41):661–690

Paterson WSB (1994) The physics of glaciers. Butterworth Heinemann, Oxford/Burlington

Providoli S, Curdy P, Elsig P (2015) 400 Jahre im Gletschereis: der Theodulpass bei Zermatt und sein "Söldner". In: Providoli S, Curdy P, Elsig P (eds) hier + jetzt, Baden

Steiner JF, Kraaijenbrink PD, Jiduc SG, Immerzeel WW (2018) Brief communication: the Khurdopin glacier surge revisited – extreme flow velocities and formation of a dammed lake in 2017. Cryosphere 12(1):95–101. https://doi.org/10.5194/tc-12-95-2018

Truffer M, Harrison WD, Echelmeyer KA (2000) Glacier motion dominated by processes deep in underlying till. J Glaciol 46:213–221

Vincent C, Moreau L (2016) Sliding velocity fluctuations and subglacial hydrology over the last two decades on Argentière glacier, Mont Blanc area. J Glaciol 62(235):805–815. https://doi.org/10.1017/jog.2016.35

Weertmann J (1964) The theory of glacier sliding. J Glaciol 5:287–303

Wendt A, Mayer C, Lambrecht A, Floricioiu D (2017) A glacier surge of Bivachny Glacier, Pamir Mountains, observed by a time series of high-resolution digital elevation models and glacier velocities. Remote Sens 9(4):388. https://doi.org/10.3390/rs9040388

Mass and Energy Balance of Glaciers

Contents

© The Author(s), under exclusive license to Springer-Verlag
GmbH, DE, part of Springer Nature 2022
W. Hagg, *Glaciology and Glacial Geomorphology*,
https://doi.org/10.1007/978-3-662-64714-1_4

4

Overview

Mass changes are the most important quantity in the assessment of glacier behaviour, but they are not as easy to determine as changes in length or area. Over the last 75 years, a binding catalogue of terms and methods has emerged, all of which have advantages and disadvantages. In addition to the three classical methods of mass balance determination, a new approach has been added in recent years in the form of the gravimetric method, which has yet to establish itself in operational use, but which may open up new possibilities and opportunities in the future. Part of the mass loss is directly controlled by the energy exchange between the glacier surface and the atmosphere. This is where it is decided whether ice changes to the liquid or gaseous state and is thus lost to the glacier, and with what efficiency these phase transitions take place.

4.1 Glacier Mass Balance

Glacier mass balance is the change in glacier mass over a given period of time. As with any balance, a weighing is made according to the model of the beam balance (from the Latin *bilancia*, from *bi* for "double" and *lanx* for "shell"), here between processes that increase the mass of a glacier (**accumulation**) and those that cause a loss of mass (**ablation**). Depending on which scale predominates, this results in a mass gain or a mass loss for the glacier as a whole, the so-called **mass balance** or **net balance** (◘ Fig. 4.1).

The glacier mass balance is particularly important for the climatic interpretation of glacier behaviour because it represents the immediate and unfiltered

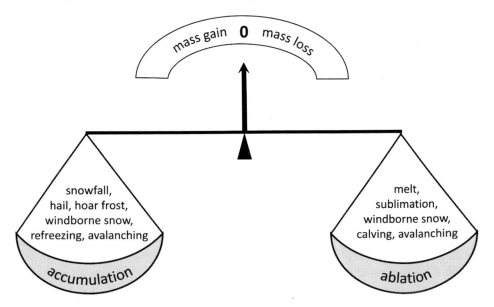

◘ **Fig. 4.1** The principle of the glacier mass balance, illustrated by means of a beam balance

response of the glacier to the local climate and is not, like glacier length or area, delayed in time and influenced by non-climatic factors (▶ Chap. 6). This makes mass balance the most important metric for assessing glacier behavior and for inferring climate variability. In this chapter, the concept and components of the glacier mass balance are first described, before measurement methods are explained and, at the end, the worldwide distribution of mass balance data is briefly considered.

❯ The glacier mass balance is the difference between all mass gains, which are called accumulation, and all mass losses, which can be summarized under the term ablation.

4.1.1 Concept and Components of the Glacier Mass Balance

As already shown in ▶ Chap. 2, glacial ice is formed from the metamorphosis of snow, which is called firn once it has survived a full summer. At a certain thickness, the ice begins to deform under pressure (▶ Chap. 3) and, following gravity, flows toward the valley, where melt rates increase and winter snowfall decreases. The mass change of a glacier is usually given over a year and, analogous to precipitation data, is expressed in terms of water equivalent. This is the height of the water column that would result from melting. Common units are millimetres, centimetres or metres of water equivalent per year (mm, cm, m w.e. a^{-1}). The glacier mass is understood as the total mass of all components of the glacier and includes not only the masses of ice, firn and snow but also those of liquid water and inclusions or overlays of rock (englacial and supraglacial moraine; ▶ Chap. 10).

Accumulation includes not only snowfall but also deposition by hail, hoar frost, windborne snow, avalanches and refreezing rain or meltwater. Snowfall is usually the most significant process in quantitative terms, but in cirques or with regenerated glaciers (▶ Chap. 5), snow or ice avalanches can also account for the main part of the accumulation.

In addition to meltwater runoff, **ablation** is also composed of sublimation, wind drift, calving and ice avalanches. In mountain glaciers, melting is the most important process of ablation; exceptions are steep and high-elevation hanging glaciers, which lose the most mass through ice avalanches. For glaciers that flow into lakes or sea bays, calving processes may dominate ablation. This is the breaking off of smaller or larger icebergs at the glacier front, which is also referred to here as the calving front and is often formed as a steep and sometimes several tens of metres high ice cliff.

Most of the mass turnover takes place on the glacier surface. Internal accumulation by refreezing and internal ablation by melt in crevasses and channels is already significantly lower than at the surface, and basal accumulation and ablation are negligible except for subaqueous melt of floating glacier tongues.

Since in most glaciated areas accumulation and ablation peak at different times of the year, the mass of glaciers varies throughout the year. In the mid-latitudes of the northern hemisphere, glaciers have their greatest mass at the end of the main

4

winter accumulation period and their least mass at the end of the autumn melt period. The natural glaciological **budget year** is variable and extends over two mass minima, i.e. from autumn to fall in the example given. However, the exact time varies from year to year depending on the weather; it is usually difficult to match this time for elaborate measurement campaigns that cannot be carried out several times for logistical or financial reasons. For this reason, most mass balance measurements use the glaciological budget year (also known as the **fixed date system**), which in the northern hemisphere begins on 1st October and ends on 30th September. Around this date, the snow accumulation is usually lowest, after which the build-up of the winter snow cover begins. Large deviations from the natural budget year only occur in the case of particularly early or late onsets of winter; the error resulting from this shift is then accepted.

As mentioned earlier, accumulation and ablation on most glaciers occur largely separately seasonally, namely accumulation in winter and ablation in summer. This seasonal mass balance cycle is termed **winter-accumulation type** (Ageta and Higuchi 1984) and is typical of most extratropical glacier regions. The time course of accumulation and ablation for this type is shown in ◘ Fig. 4.2.

After an autumnal minimum (time z_1), the **accumulation period** begins with the first snowfall, which will not melt anymore in the same year. Since ablation is negligible in winter, the mass balance curve up to time z_w the maximum of snow accumulation, is essentially the same as that of accumulation. The mass balance at time z_w (for the fixed date system: 30th April) corresponds to the winter balance; thereafter the **ablation period** begins, which lasts until the next autumnal minimum at time z_2. Since there is little accumulation due to snowfall in summer, the shape of

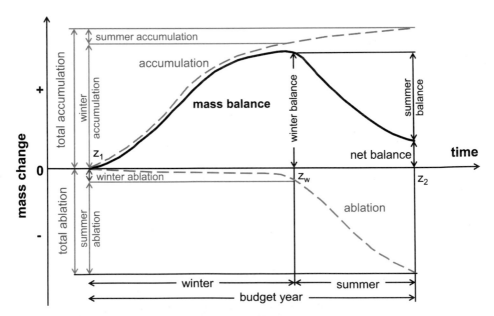

◘ **Fig. 4.2** Annual course of mass balance for a glacier with predominant winter accumulation. (Modified after Paterson 1994; with kind permission of © Elsevier AG 1994, all rights reserved)

the mass balance curve in this half-year largely corresponds to that of ablation. The mass balance at time z_2 corresponds to the annual balance and is also referred to as the net balance. If the winter balance is determined by measurements at time z_w, then the summer balance can also be determined from the net balance minus the winter balance. This is very advantageous for the climatic interpretation of the glacier mass balance (▶ Chap. 6). If, on the other hand, only one measurement campaign takes place at time z_2, only the annual balance can be given for this budget year.

On tropical glaciers, which are mainly found in South America and in monsoon Asia, accumulation and ablation peak simultaneously. The annual course of these **summer-accumulation type** glaciers (Ageta and Higuchi 1984) is shown in ◘ Fig. 4.3.

Both mass gains and mass losses occur throughout the year, but in the example shown both processes are strongest in summer. Since ablation predominates in ◘ Fig. 4.3, this results in a negative net balance. This glacier nourishment type is particularly sensitive to climate variability because warming of air temperature here affects both scales of the mass balance: Increased ablation in the warmer atmosphere is compounded by a reduction in accumulation because less precipitation falls in solid form.

The dominance of accumulation and ablation at different altitudes creates a spatial differentiation of the glacier surface with regard to the mass balance. In the warmest, lowest areas, the snow cover, which is also less thick here than higher up, melts out first in spring. This process leaves bare ice surfaces on the glacier, which are also subject to ablation after the snow has melted. The temporary snow line, i.e. the lower limit of the current vertical snow distribution, moves to higher and higher altitudes over the summer. This process can be briefly interrupted by summer snowfalls, which immediately stop ice melt (▶ Chap. 7). Since such summer snowfalls are usually not very productive, they quickly disappear under the summer sun, and the original snow-ice pattern is soon restored. Before the first autumnal snows, which are not melted again in the same year, the temporary snow line

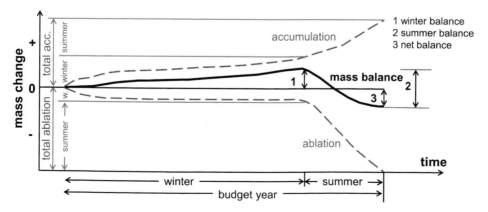

◘ **Fig. 4.3** Annual course of mass balance for a glacier with predominant summer accumulation. (Modified after Ageta and Higuchi 1984)

4

then reaches a maximum height. Above this height, called **equilibrium line altitude (ELA),** remnants of snow from the previous winter are still preserved, indicating that the glacier has experienced mass gain on these surfaces. In ◨ Fig. 4.4, the equilibrium line is higher than in the previous year, which is why a firn zone exists here between the snow line and the snow-free ice. In years when the equilibrium line is lower than in previous years, only snow and ice surfaces exist. The equilibrium line is always identical with the snow line, although this term is not unproblematic here because snow also becomes firn by definition at the end of summer. Strictly speaking, one would have to speak of new and old firn, but for the sake of simplicity, the snow from last winter is still referred to as such here. Below the equilibrium line, either firn and ice melt or only ice melt has taken place; in any case, the glacier has lost mass in the year under consideration. The two areas with mass gain and mass loss, which are decisive for the mass balance, are called the **accumulation area** and the **ablation area.** Directly on the equilibrium line, the net balance is zero, accumulation and ablation are exactly balanced here (and only here).

The proportion of the accumulation area to the total glacier area is referred to as the **accumulation area ratio (AAR)** and corresponds to the proportion of snow cover at the end of the ablation period (time z_2 in ◨ Fig. 4.2). For stationary glaciers in equilibrium with a stable climate, the AAR averages 67%, which means that two thirds of the glacier area is still covered with snow in autumn.

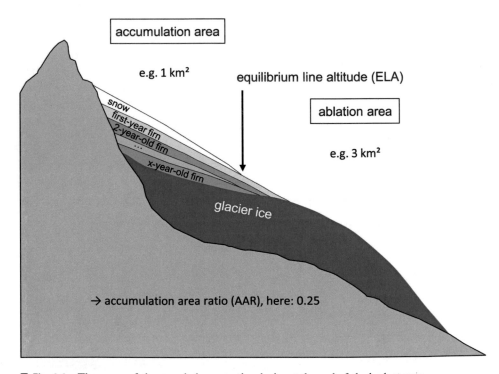

accumulation area

e.g. 1 km²

equilibrium line altitude (ELA)

ablation area

snow
first-year firn
2-year-old firn
…
x-year-old firn

e.g. 3 km²

glacier ice

→ accumulation area ratio (AAR), here: 0.25

◨ **Fig. 4.4** The zones of the mass balance on the glacier at the end of the budget year

❯ The upper area of a glacier where mass gains are achieved is called accumulation area and is separated by the equilibrium line from the ablation area where mass losses are recorded. The equilibrium line corresponds to the snow line at the end of the ablation period for glaciers with predominant winter accumulation.

4.1.2 Methods of Mass Balance Determination

The first mass balance measurement was made on the Storglaciären in Sweden in 1945. Here, accumulation and ablation were measured directly on the glacier surface, which is why the method is also called the direct, glaciological or traditional method. Advances in surveying and remote sensing allowed the development and improvement of the so-called geodetic method, where the glacier does not necessarily have to be stepped on. For the sake of completeness, the hydrological-meteorological method should also be mentioned here, although it is hardly used due to its high measurement effort and high uncertainties in the determination of water balance terms. As a completely new and promising approach for the future, gravimetry has also been used in recent years to determine mass changes. All four methods are described below with regard to their approach, advantages and limitations.

In the **glaciological method**, described in detail by Østrem and Brugman (1991), the two terms of the mass balance are measured directly on the glacier. For this purpose, ablation stakes are drilled into the ice in the ablation area at the beginning of a budget year, i.e. at the time of minimum snowcover (or – with the *fixed date system* – at 1st October). These poles often consist of individual segments, and since they must not melt under any circumstances, they must either be stuck in the ice at least as deep as the maximum expected melt in the following summer, or they must be re-drilled during the melt period. At low-lying glacier tongues, annual ablation can be 8–10 m, and such deep holes are usually melted into the ice with steam-powered drills. The visible length of the stake above the ice surface is measuredand the reading is repeated after 1 year; the difference between the two measurements gives the lowering of the ice surface at that location (❑ Fig. 4.5b). Multiplied by the density of ice (0.9 g cm^{-3}), this value becomes the water equivalent or mass balance point value.

At the end of the budget year, it is also necessary to determine the accumulation, i.e. the mass of snow that has survived the summer in the accumulation area. For this purpose, snow pits are dug down to the autumn horizon of the previous year, which can usually be clearly identified as a hard layer, resulting from melting and refreezing processes during the summer. Sometimes this layer boundary is also readily identifiable as a dirt or dust layer, but it can also be artificially marked by the application of adye tracer. At the snow pits it is not sufficient to measure the height of the snow layer, but here too the water equivalent must be determined. To do this, the entire snow pack is punctured from top to bottom with cylinders and each puncture is weighed with a spring balance (❑ Fig. 4.5a). From the mass and volume, the density of the respective layer can be determined and thus the water equivalent for the layer and finally for the entire snowpack can be calculated. Since the amount of work

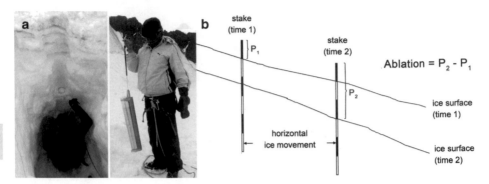

⬛ Fig. 4.5 Glaciological method of mass balance determination. (**a**) Determination of accumulation by cutting out snow volumes with a cylinder (left, cylinder in centre of picture) and weighing it (right) on Golubin Glacier, Kyrgyzstan. (Photos: W. Hagg); (**b**) Schematic of ablation measurement with a stake

required for snow pits is relatively high and the density of the snow does not show a very large spatial variation, usually only a few pits are dug and between them numerous quickly feasible snow thickness measurements are made with thin metal probes. However, this only works well when the fall horizon is clearly formed as a hard ice layer. All water equivalents – the positive ones from the accumulation area and the ones with a negative sign from the ablation area – are located as point measurements on a map or a Geographical Information System and intra – and extrapolated to the entire glacier area. It is assumed that the measured values change continuously between two points (intrapolation) and also beyond (extrapolation) or according to a certain mathematical pattern. In this step lies the greatest uncertainty of the glaciological method (Zemp et al. 2013). Often not all glacier areas can be reached with reasonable effort and safety, and especially the marginal areas may show larger variations in the mass balance due to snow displacements caused by wind and avalanches or by shading effects of rock walls, so that larger errors may occur by extrapolation. However, an undoubted advantage of the glaciological method is that annual mass balance values and even winter and summer balances can be produced if a second survey is carried out at the end of the accumulation period. In addition, mass balances can be determined spatially distributed for individual glacier areas.

> The glaciological method determines accumulation and ablation directly on the glacier. This allows annual and even seasonal mass balances to be determined. The greatest source of error lies in the transfer of point measurements to the glacier as a whole, especially if larger areas have not been sampled.

The **geodetic method** is based on the comparison of the glacier surface height at two different points in time. The surface elevation can be determined using various methods of earth surveying (geodesy). The first accurate glacier maps have existed since the end of the nineteenth century, initially produced using photogrammetry (Finsterwalder 1897). This method is based on taking photographic images from different perspectives. If the coordinates of the photo viewpoints can be determined by other methods (e.g. taking bearings on known mountain peaks), then coordinates of points identifiable in two photos with different perspectives can be determined using triangular geometry. Since the mid-twentieth century, also aerial photographs have been evaluated photogrammetrically; towards the end of the century, optical satellite data were added (Kaab 2002; Bolch et al. 2008). The analysis is now done digitally, which has automated many labor-intensive steps such as control point generation. This millennium has seen terrestrial and airborne laser scanning (Geist et al. 2005), global navigation satellite systems (Hagen et al. 2005), and the use of drones (Bhardwaj et al. 2016) emerge as cost-effective and increasingly accurate alternatives.

By subtracting the surface elevations, the height difference between the two times can be determined (● Fig. 4.6). Assuming an average glacier density, the mass balance can be calculated from this. In addition to the uncertainties of the measurement methods for determining surface elevations, the assumption of a mean density is the largest source of error in the geodetic method. Since mass change occurs not only in ice, but also to (usually unknown) proportions in firn and snow, a mixed density is used as a basis. Huss (2013) proposes a value of $0.85 \, \mathrm{g \, cm^{-3}}$.

An advantage, at least of some surveying methods, is the fact that the glacier does not have to be stepped on, which makes it possible to record glaciers or glacier areas that are difficult to access. In addition, some methods do not provide point measurements, but rather areal information and are thus not subject to interpolation error. With precise measurement methods, the result can be more accurate than with the glaciological method. However, these very accurate methods such as airborne laser scanning (ALS) are often too expensive for annual use. Methods with larger errors, such as photogrammetric surveys, used to be applied mostly only over periods of at least 10 years, but major technical advances have recently taken place here, so that mass balances can now also be determined over shorter periods (● Fig. 4.6).

Since the geodetic method does not take ice movement into account, mass balance values can only be given for the entire glacier and not for sub-areas as with the glaciological method. This is most obvious if one imagines a stationary glacier that does not change its surface. The geodetic method would give a mass balance of zero for each point, but this does not correspond to reality: the glaciological method would indicate a mass gain in the accumulation area and a mass loss in the ablation area, both of which actually take place in a glacier in equilibrium and are compensated for by ice movement. Only the mean value would also be zero with the glaciological method and would agree with the result of the geodetic method. The geodetic method is well suited for checking annual balance values of the glaciological method and for eliminating any systematic errors in the annual values.

> ❯ In the geodetic method, the height of the glacier surface is measured twice and from this the volume difference is determined, from which the mass change can be calculated. To do this, one must assume an average density, which can lead to errors. With many geodetic measurement methods, the ice does not have to be stepped on, which is an advantage for large glaciers or areas that are difficult to access.

The **hydrological-meteorological method** assumes that hydrological storage changes in high mountains consist mainly of glacier mass changes. By solving the water

4

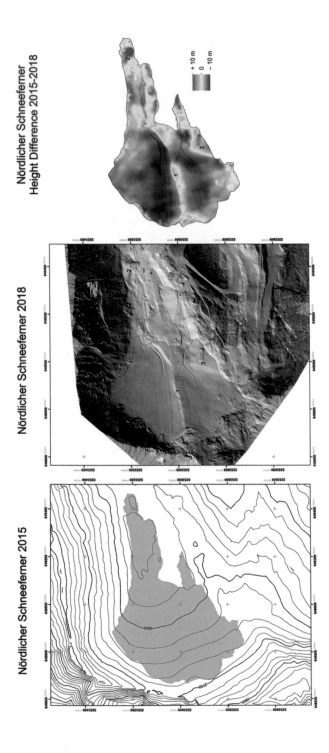

■ **Fig. 4.6** Application of the geodetic method. The digital elevation models are visualized once as contour plot (2015) and once as hillshade; they were produced with GPS measurements (2015) and with digital airborne photogrammetry (2018). The difference grid (2015–2018) shows areas of uplift (bluish colors) in addition to predominant surface lowering (reddish colors). These can all be explained by activities of ski area operators: Snow redistributions are used to secure a drag lift route (1), to create snow depots for autumn (2) or to prepare an igloo village (3). The mean value of the height change is −3.11 m or −1.04 m per year. As the mass changes here occur exclusively in the ice, a mean density of 0.9 g cm^{-3} can be assumed; the mean annual mass balance in the period 2015–2018 is therefore −0.93 m w.e. (Data source: ▶ www.bayerische-gletscher.de)

balance equation for the storage changes (ΔS) (Eq. 4.1) and equating these with glacier mass changes, we obtain:

$$MB\,(\Delta S) = N - V - A \tag{4.1}$$

- MB = Glacier mass balance
- N = Precipitation
- V = Evaporation
- A = Runoff

The mass balance calculated in this way is initially related to the entire area of the hydrological catchment. This is the area from which water flows to the runoff gauge; this also includes unglaciated areas. Even if the discharge were measured directly at the glacier gate, there are usually still rock and summit areas in the catchment that are not ice-covered. The farther the gauge is from the glacier, the larger the ice-free portions become. Therefore, to convert the mass balance to the glacier area, the proportion of glacierization must be taken into account. If the catchment area contains several glaciers, the balance is to be considered as an average value and cannot be distributed spatially.

The glacier does not have to be entered in this procedure and large glaciers or inaccessible glacier areas can also be recorded. If storage changes of lakes, groundwater and snow cover exist, these will lead to errors if they are not determined and taken into account separately. However, these storage changes over a budget year are often so small compared to the glacier mass balance that they can be neglected.

A major disadvantage is the logistical effort and uncertainties in determining precipitation, evaporation and runoff. Continuous measurement of sediment-laden glacial runoff requires not only suitable instrumentation but also a great deal of maintenance and repeated calibration measurements, because the stream cross-section and thus the stage-discharge relationship are changing constantly. The monitoring of actual evaporation is not possible for a high mountain catchment, but the value is low compared to the other terms of the water balance and can be estimated well with empirical formulas. The greatest uncertainty is associated with precipitation measurement, which is generally subject to error, for example because the measuring device represents a disturbance in the wind field and the amount of rain or snow collected is not necessarily representative. Especially in high mountains with frequent snowfall and simultaneous wind, a precipitation collector can behave like a moving car whose windshield never comes into contact with snowflakes. Since the strong relief results in pronounced spatial differences in precipitation distribution, for example due to elevation or windward/leeward effects, the interpolation of point measurements and thus the determination of the mean precipitation in the catchment area (the so-called basin precipitation) is often subject to a large error. Since precipitation takes the largest value in the water balance equation, the error may even be of the order of magnitude of the sought term, i.e. the glacier mass balance, which is of course extremely unfavourable. On the other hand, the method is the only one that can account for changes in liquid water accumulations in the glacier (Tangborn et al. 1975; Paterson 1994).

> The hydrological-meteorological method calculates the mass balance as a residual element of the water balance equation from the mean precipitation minus evaporation and runoff. The glacier does not have to be entered for this, but difficulties in determining the basin precipitation can lead to large errors, and the permanent measurement of runoff is very laborious.

The range of methods for determining the mass balance has been extended in recent years to include the **gravimetric method.** The change in the Earth's gravitational field is directly dependent on the redistribution of mass in the Earth system. Gravimetric methods can be used to measure the gravitational field and its changes and thus draw conclusions about changes in mass. The GRACE (Gravity Recovery and Climate Experiment) satellite system, which has been in orbit since 2002, consists of two satellites that continuously and very precisely determine their distance from each other. If a satellite flies over a region with higher gravity (mass), it accelerates and the distance to the other satellite increases. However, the spatial resolution of gravity field and mass changes determined by this principle is many hundreds of kilometers and can only be applied to ice sheets (Velicogna and Wahr 2005) and extensive glacial regions such as Alaska (Luthcke et al. 2008) or the Tienshan Mountains (Farinotti et al. 2015). The challenge in data evaluation lies in splitting the signal into its individual components, because all mass changes, e.g. also hydrological ones, are included here.

Individual mountain glaciers can therefore not yet be recorded with satellite gravimetry; this is only possible with terrestrial gravimetry, which can register mass changes corresponding to a melting of the glacier surface of a few decimetres. However, this method is very complex and is therefore so far only used in special measurement programmes, but not in routine operations (Gerlach 2013). Flight gravimetry is more effective than terrestrial measurements. The resolution here is a few kilometres, but the accuracy is also two to three orders of magnitude lower, making it unsuitable for glacier mass balances, at least over shorter periods of time (Gerlach 2013).

The disadvantages of gravimetric methods are the effort and uncertainties in data evaluation and spatial resolution. In the future, gravimetry, especially satellite – or aircraft-based, could develop into an additional, independent method of mass balance determination.

> Mass changes affect the Earth's gravitational field, which are measurable and allow conclusions to be drawn about changes in the mass distribution. However, so far only large glacial areas can be detected with reasonable effort and measurement error using this method, and here the separation of ice masses from other mass signals can be difficult.

4.1.3 **Mass Balance Measurements Worldwide**

Systematic records of glacier changes began in 1894 with the founding of the *Commission Internationale des Glaciers* in Zurich. Initially, these were mainly changes in length, but it was not until 1945 that mass balance measurements were

▢ **Table. 4.1** Glacier area and number of glaciers observed in the 2015/2016 and 2016/2017 budget years in different macro-regions of the WGMS

	Glacier area(km²)	Number of current mass balance series	Number of reference glaciers
Alaska	86,500	4	2
Western North America	14,500	19	7
Canadian Arctic	146,000	4	4
Greenland	89,500	3	0
Iceland	11,000	9	0
Svalbard and Jan Mayen	34,000	12	2
Scandinavia	3000	14	9
Central Europe	2000	50	12
Caucasus and Middle East	1500	2	2
Russian Arctic	51,500	0	0
North Asia	2500	2	0
Central Asia	49,500	15	2
South Asia	48,500	14	0
Tropics	2500	7	0
Southern Andes	29,500	10	1
New Zealand	1000	2	0
Antarctica (incl. islands)	133,000	3	0
Worldwide	**706,000**	**170**	**40**

Source: WGMS (2021)

added using the glaciological method. Today, standardised data are collected by the World Glacier Monitoring Service (WGMS) in Zurich. The WGMS does not carry out any measurement campaigns itself, but archives and maintains all the data supplied to it by a worldwide network of national correspondents. The database, which is also accessible via the Internet site ▶ www.wgms.ch, currently comprises 48,557 length changes of 2620 glaciers and 7386 mass balance measurements of 482 glaciers. Current measurement series exist at 170 glaciers, 41 of which are longer than 30 years (WGMS 2021). These glaciers with long observation series, which are particularly suitable for climatic interpretations, are referred to as reference glaciers. ▢ Table 4.1 illustrates the imbalance between the glacier areas of individual regions and the number of mass balance measurement series. Whereas

4

in Central Europe (Alps, Pyrenees, Apennines) there are 52 current measurement series and 11 reference glaciers, in the southern Andes, for example, where the glacier area is almost 15 times as large, there is only one glacier with a measurement series of over 30 years. In South Asia, where the glaciated area is much larger, there is not a single reference glacier, and in the Russian Arctic not even a single current measurement series.

From 1991 to 2000, the mean AAR for the reference glaciers of the WGMS (2021) was 45%, from 2001 to 2010 it was 35%, and from 2011 to 2020 it was only 30%. Thus, instead of two-thirds, as would be the case for glaciers in equilibrium, less than one third of the glacier area was covered with snow at the end of summer during the past decade. This leads to continued negative mass balances, which will be discussed in more detail in ▶ Chap. 8.

4.2 Energy Balance of Glacier Surfaces

Similar to the comparison of the gain and loss of mass in the case of mass changes, one can also balance the inflow and outflow of energy. Various processes add energy to the glacier, while others take it away. The sum of these individual terms is the energy balance. If it is positive, melting is taking place. It should be emphasised at this point that only the surface energy balance at the contact with the atmosphere, is considered here. Energy exchange also takes place at the glacier bed, but this is negligible compared to the surface and can be disregarded for the energy balance of the glacier as a whole.

The most important term for energy turnover is the **net radiation**, i.e. the difference between incoming and outgoing radiation. Another source of energy is the **sensible heat flux**, i.e. thermal energy, which is expressed in an increase or decrease in air temperature. The **latent heat flux** is coupled to phase transitions of water; it can appear through condensation (energy supply) or evaporation (energy removal), thus having a positive or a negative sign. The heat input by rain and the ground heat flux into or out of the ice hardly play a role quantitatively. These two quantities can therefore be neglected, so that the simplified energy balance equation (Eq. 4.2) is as follows:

$$M = R + H + LE \tag{4.2}$$

- M = Total energy available for melting
- R = Radiation balance (net radiation)
- H = Sensible heat flux (temperature)
- LE = Latent heat flux (evaporation, condensation)

The energy fluxes through radiation must be separated into two wavelength ranges on closer examination, short-wave and long-wave radiation. **Short-wave radiation** is solar radiation and thus the primary source of energy on the earth's surface, while **long-wave radiation** refers to the thermal radiation that emanates from every body.

Incoming short-wave radiation is also referred to as global radiation (S_G). It consists of the radiation that passes through the atmosphere unhindered (**direct**

radiation) and that which is scattered by cloud components or other particles (**diffuse radiation**). Depending on the reflectivity of the surface, the so-called albedo (a), part of the incoming global radiation is reflected, the other is absorbed and converted into heat. A perfectly white body has an albedo of 100% or 1; on glaciers, the value varies roughly between 0.9 (fresh snow) and 0.2 (dark ice), and it can drop to 0.1 for debris cover (Paterson 1994). This means that between 10% (fresh snow) and 90% (debris) of the solar radiation is absorbed. The short-wave net radiation is calculated from the global radiation (S_G) multiplied by the reciprocal of the albedo (1 − a), i.e. the absorbed proportion of the irradiation.

Every body emits long-wave thermal radiation. According to the Stefan-Boltzmann law, this increases very strongly with the temperature of the body. For example, a warm oven emits more thermal radiation than a cold one. For melting ice surfaces, the temperature is 0 °C. From the emissivity of ice and the Stefan-Boltzmann constant, this results in a constant value for thermal radiation of 315 W per square metre. A part of the long-wave radiation goes into space and leaves the earth-atmosphere system, another part is reflected in the atmosphere and reaches the earth or ice surface again as incoming longwave radiation. The reflection of the outgoing long-wave radiation takes place at water vapour, other greenhouse gases and clouds. This component of the radiation balance is also the reason why nights with cloud cover are milder than those with clear skies. Thus, also in the long-wave range, there are energy fluxes in both directions: away from and towards the glacier surface. These are offset against each other in the long-wave net radiation.

The shortwave and longwave balances can be combined in the total net radiation R (Eq. 4.3):

$$R = S_G \left(1-a\right)+\varepsilon\ \sigma\ T^4 +L_G \tag{4.3}$$

- S_G = Global radiation [Wm^{-2}] = direct + diffuse radiation
- a = Albedo = proportion of short-wave reflection
- ε = Emissivity of ice (0,98 … 1,0)
- σ = Stefan-Boltzmann constant $5.6697^{-}10^{-8}$ ($Wm^{-2}\ K^{-4}$)
- T = Surface temperature (K)
- L_G = Long-wave counter radiation (Wm^{-2})

The net radiation is the most important source of energy for the ice surface. On alpine glaciers it accounts for about 75–80% of the total heat gain during the main ablation period (July, August). In second place is the sensible heat flux with an average of about 15–20%. Evaporation and condensation can be important short-term energy sources or sinks. An important control factor in this context is humidity. If the pressure of the water vapour, i.e. the gaseous water in the air, exceeds a value of 6.1 hectopascals (hPa), then condensation occurs above melting ice surfaces and thus energy is gained; at lower values, evaporation conditions prevail. Because evaporation of melted ice that has already been lost to the glacier consumes a lot of energy, less is left for further melting. So dry air, by providing evaporative conditions, reduces ablation. In the long run, however, the two processes more or less balance each other out in the Alpine region, so that latent heat fluxes are shortened

4

◘ Fig. 4.7   Snow penitents in the Elbrus region, Caucasus (left) and in the Fedchenko region, Pamir (right). (Photo left: W. Hagg, photo right: Ludwig Braun)

from the equation or, if condensation slightly predominates, represent a small energy gain. The total contribution of latent heat to the total energy gain is about 0–10% in the Alpine region. These figures were derived from elaborate short-term experiments (Hoinkes 1953; Funk 1985; Moser et al. 1986; Greuell and Oerlemans 1987; Weber 2008) and provide only rough estimates for summer conditions on alpine glaciers; over short periods, however, they are subject to large, weather-related fluctuations.

In other climates, the importance of the individual terms of the energy balance can deviate significantly from the alpine conditions. In continental, dry climates, the relative importance of radiation increases even more with the decrease in cloud cover; with lower humidity, evaporation is promoted here at the same time. This can mean a considerable loss of energy for the ice surfaces. In the case of strong shortwave radiation and dry air, sublimation can even occur to a greater extent; during this phase transition, 8.5 times as much energy is consumed as during melting. In the case of strong sublimation, centimetre to metre-sized jagged shapes can develop from depressions in the snow and ice. While sublimation occurs in dry air at the tips of the jags, melting processes can still occur in the more wind-protected and moist depressions in between. Because melting is more efficient, the jags become larger and larger as long as the sun's rays still reach the depressions. These forms also occur in subtropical arid regions and are called **snow penitents** (◘ Fig. 4.7).

In maritime climates, there is a decrease in short-wave solar radiation with heavier cloud cover. Even if the long-wave incoming radiation increases, the net radiation as a whole loses significance for energy gain. At the same time, latent heat can contribute significantly to melt energy on these glaciers, where condensation conditions clearly predominate (Winkler 2009).

> The largest source of energy for most glaciers is solar radiation, followed by sensible heat. Latent heat fluxes represent a loss of energy under evaporative conditions in dry air, this effect predominates in continental climates. At high humidity on maritime glaciers, condensation heat can contribute significantly to melt energy.

References

Ageta Y, Higuchi K (1984) Estimation of mass balance components of a summer-accumulation type glacier in the Nepal Himalaya. Geogra Ann Ser A Phys Geogr 66:249–255. https://doi.org/10.2307/520698

Bhardwaj A, Sam L, Martín-Torres FJ, Kumar R (2016) Remote sensing of 427 environment UAVs as remote sensing platform in glaciology: present applications and future 428 prospects. Remote Sens Environ 175:196–204. https://doi.org/10.1016/j.rse.2015.12.029

Bolch T, Buchroithner MF, Pieczonka T, Kunert A (2008) Planimetric and volumetric glacier changes in the Khumbu Himal, Nepal, since 1962 using Corona, Landsat TM and ASTER data. J Glaciol 54(187):592–600

Farinotti D, Longuevergne L, Moholdt G, Duethmann D, Molg T, Bolch T, Vorogushyn S, Guntner A (2015) Substantial glacier mass loss in the Tien Shan over the past 50 years. Nat Geosci 8:716–722. https://doi.org/10.1038/ngeo2513

Finsterwalder S (1897) Der Vernagtferner – seine Geschichte und seine Vermessung in den Jahren 1888 und 1889. Wiss Ergänzungshefte Z Dtsch Oesterr Alpenvereins 1(1), Verlag des DuÖAV, Graz

Funk M (1985) Räumliche Verteilung der Massenbilanz auf dem Rhonegletscher und ihre Beziehung zu Klimaelementen. Zürcher Geogr Schriften 24:183

Geist T, Elvehøy H, Jackson M, Stotter J (2005) Investigations on intra-annual elevation changes using multitemporal airborne laser scanning data – case study Engabreen, Norway. Ann Glaciol 42:195–201

Gerlach C (2013) Gravimetrie und deren Potential für eine unabhängige Bestimmung der Massenbilanz des Vernagtferners. Z Gletscherk Glazialgeol 45(46):281–293

Greuell W, Oerlemans J (1987) Energy balance calculations on and near Hintereisferner (Austria) and an estimate of the effect of greenhouse warming on ablation. In: Oerlemans J Glacier fluctuations and climatic change. Glaciology and quaternary geology, Kluwer Academic Publishers, Dordrecht, S 305–S 323

Hagen JO, Eiken T, Kohler J, Melvold K (2005) Geometry changes on Svalbard glaciers: mass-balance or dynamic response? Ann Glaciol 42:255–261

Hoinkes HC (1953) Zur Mikrometeorologie der eisnahen Luftschicht. Arch Met Geoph Biokl B2:451–465

Huss M (2013) Density assumptions for converting geodetic glacier volume change to mass change. Cryosphere 7:877–887. https://doi.org/10.5194/tc-7-877-2013

Kaab A (2002) Monitoring high-mountain terrain deformation from repeated air- and spaceborne optical data: examples using digital aerial imagery and ASTER data. ISPRS J Photogramm Remote Sens 57(1–2):39–52

Luthcke SB, Arendt A, Rowlands D, McCarthy J, Larsen CF (2008) Recent glacier mass changes in the Gulf of Alaska region from GRACE mascon solutions. J Glaciol 54(188):767–777. https://doi.org/10.3189/002214308787779933

Moser H, Escher-Vetter H, Oerter H, Reinwarth O, Zunke D (1986) Abfluß in und von Gletschern. GSF-Bericht 41, Teil I u. II. GSF Gesellschaft für Strahlen- und Umweltforschung, München

Østrem G, Brugman M (1991) Glacier mass-balance measurements: a manual for field and office work. NHRI science report. National Hydrology Research Institute, Saskatoon

Paterson WSB (1994) The physics of glaciers. Butterworth Heinemann, Oxford/Burlington

Tangborn WV, Krimmel RM, Meier MF (1975) A comparison of glacier mass balance by glacier hydrology and mapping methods, South Cascade Glacier. In: Snow and ice symposium – Neiges et Glaces. Proceedings of the Moscow symposium, August 1971. IAHS publication 104. International Association of Hydrological Sciences, Washington, DC, pp S 185–S 196

Velicogna I, Wahr J (2005) Greenland mass balance from GRACE. Geophys Res Lett 32:L14501. https://doi.org/10.1029/2005GL023458

Weber M (2008) Mikrometeorologische Prozesse bei der Ablation eines Alpengletschers. Abhandlungen. Verl. der Bayerischen Akad. der Wiss., München, S 177. ISBN 978-3-7696-2564-6 http://publikationen.badw.dc/de/013223297

WGMS (2021) In: Global glacier change bulletin no. 4 (2018–2019), Zemp M, Nussbaumer SU, Gärtner-Roer I, Bannwart J, Paul F, Hoelzle M (eds) ISC(WDS)/IUGG(IACS)/UNEP/UNESCO/WMO. World Glacier Monitoring Service, Zurich, p S 278, publication based on database version. https://doi.org/10.5904/wgms-fog-2021-05

Winkler (2009) Gletscher und ihre Landschaften. WBG, Darmstadt

Zemp M, Thibert E, Huss M, Stumm D, Rolstad Denby C, Nuth C, Nussbaumer SU, Moholdt G, Mercer A, Mayer C, Joerg PC, Jansson P, Hynek B, Fischer A, Escher-Vetter H, Elvehøy H, Andreassen LM (2013) Reanalysing glacier mass balance measurement series. Cryosphere 7:1227–1245. https://doi.org/10.5194/tc-7-1227-2013

4

Glacier Types and Distribution

Contents

© The Author(s), under exclusive license to Springer-Verlag
GmbH, DE, part of Springer Nature 2022
W. Hagg, *Glaciology and Glacial Geomorphology*,
https://doi.org/10.1007/978-3-662-64714-1_5

Overview
There are several approaches to pigeonholing the diversity of glaciers, and not all of them are very happy. In classical German literature, there is a typification according to the mode of feeding, which is hardly in use today. The most commonly used classification is morphological glacier types, where, strictly speaking, glaciers are categorized according to their size, their position in the relief and its influence on ice movement. The subdivision according to thermal criteria into temperate, cold and polythermal glaciers originates from geophysics. Today, there are slightly more than 200,000 glaciers worldwide with a total area of about 700,000 km², not including the two ice sheets of Greenland and Antarctica. Most of these glaciers are comparatively small, but much of the area is contained in the few large glaciers that are found mainly in regions closer to the poles. Iceland's largest glacier, for example, is more than four times the size of all Alpine glaciers combined.

5

5.1 Typification of Glaciers

Different approaches to classifying glaciers led to some confusion of terms in the first half of the twentieth century (Schneider 1962). Classifications were made according to geographical distribution (e.g. Himalayan or Pyrenean type), according to climatic zone (e.g. tropical or subtropical type) or according to size classes (e.g. inland ice, ice cap). Apart from the general problem of cataloguing complex natural phenomena, problems of delimitation and overlap also arise from the differences in the above principles of classification. Nevertheless, many of the older terms can still be found in the current literature, which is why three important classification systems will be presented here in their main features. While in the classical literature the mode of nourishment (Schneider 1962) and the morphology (von Klebelsberg 1948) were often used as main criteria, a typification according to thermal properties (Ahlmann 1935) seems appropriate when the movement of the ice and its geomorphological effects are the focus of consideration.

5.1.1 Typification According to the Source of Nourishment

This classification goes back to Visser (1934) and is based on the positional relationship between equilibrium line, shape and size of the glacier. In the English-speaking world, this subdivision has not become established. In the case of the *zentrale Firnhaube* ("central firn cap"), several glacier tongues flow radially in different directions from an accumulationarea. The equilibrium line is therefore more or less circular, or at least closed. This type of glacier occurs mainly in polar and subpolar regions (e.g. Mýrdalsjökull, Iceland). However, due to the rise in the climatic snowline in the course of current climate warming, more and more glacier systems in more temperate latitudes strictly speaking belong to this type. Whereas the equilibrium line used to run across the glacier tongues in former times and was interrupted in between, today it lies closed in a circle in the central firn area (e.g. Adamello Glacier, Italy).

A *Firnmuldengletscher* ("firn hollow glacier") exists when the accumulation area is in a flat form and nourishment is mainly by snowfall rather than avalanches. The glacier itself may remain confined to the hollow or, in the case of larger flattenings, form a tongue. If ice streams from other firn depressions join, this is also called a compound glacier (e.g. Aletsch Glacier, Switzerland).

If not only high-level flattenings belong to the accumulation area, but also the upper parts of the central glacier tongue, then it is a *Firnstromgletscher* ("firn stream glacier") ( Fig. 5.1).

If the equilibrium line sinks even further and the glaciers swell more, they flow into neighbouring valleys via so-called transfluence passes. When they connect with glacier branches there, one speaks of an *Eisstromnetz* (dendritic glacier system). As glaciation increases, mountain ridges slowly drown under the ice and only high peaks rise out of the sea of ice as rock islands called nunatak. Dentritic glacier systems (or "ice fields") only exist in polar regions (e.g. Alaska, Spitsbergen), but during the Pleistocene glaciations (▶ Chap. 8) the Alps were also covered by this type of glacier.

In the case of lower-lying firn basins that lie below the climatic snow line, avalanche input plays a decisive role for nourishment in addition to snowfall. In this case, the composite *Firnmuldengletscher* ("firn hollow glacier") type becomes a *Firnkesselgletscher* ("firn kettle glacier") and the singular type a *Lawinenkesselgletscher* ("avalanche kettle glacier") (■ Fig. 5.2). In the case of very strong avalanche input from very high rock outcrops, the glacier need not be confined to the shallow basin, but can also form long tongues (e.g. Shispar Glacier, Pakistan).

Flankenvereisung ("flank glaciation") occurs when the firn area is not in a flattening but in steep terrain, without ice avalanches already being the preferred type of ablation. This type of glacier is often much wider than long. If the slope is somewhat less steep and the ice is thicker, there is also talk of *Wandvergletscherung* ("wall glaciation") (Schneider 1962).

A special case of nourishment occurs when a glacier is interrupted by a steep rock wall. The lower part often lies below the climatic snowline and is then fed exclusively by ice avalanches. If it retains all the dynamic characteristics of a flowing glacier, it is called a **reconstituted glacier** (■ Fig. 5.3).

■ Fig. 5.1 Fedchenko glacier in Tajikistan. On the longest valley glacier in the world (approx. 70 km), the upper part of the glacier tongue also belongs to the accumulation zone. It is therefore a *Firnstromgletscher*. (Photo: Surat Toimastov†)

5

◘ Fig. 5.2 *Lawinenkesselgletscher* in the Alay Range, Kyrgyzstan. (Photo: W. Hagg)

◘ Fig. 5.3 Regenerated glacier: Fellaria, Italy. (Photo: Anne Nowottnick)

5.1.2 **Morphological Glacier Types**

This classification is sometimes considered problematic because it is partly purely descriptive and does not subdivide glaciers according to dynamic or process-based criteria. This results in a certain arbitrariness and difficulties in delimitation. Thus, although it does not satisfy throughout, the classification by relief type is so strongly established in the scientific and alpinist literature that it will be discussed here, at least in a rudimentary way. The basic classification results from the distinction whether the subglacial relief determines the flow direction of the ice or not.

5.1.2.1 **Unconstrained Glaciers**

Unconstrained glaciers cover the landscape so completely that the flow direction of the ice is decoupled from the underlying bedrock topography and is controlled only by the inclination of the ice surface. They are therefore unrestricted by the relief. Unconstrained glaciers include ice sheets, ice caps and piedmont glaciers.

Ice sheets are flat, slightly bulged ice masses of continental scale. The only ice sheets that exist today are the inland ice sheets of Greenland (about 1.7 million km^2) and Antarctica (about 12.3 million km^2). They either border directly on the sea, where they experience most of their mass loss through calving processes, or reach it via **outlet glaciers**. These are flow sections along valley flanks that protrude from the ice surface at the thinner margins of the inland ice. Because the flow direction of the glaciers here is again controlled by the bedrock, these parts of the ice sheets already belong to the constrained glacier type. This circumstance illustrates another problem of morphological glacier types, namely that a large glacier in different areas may be assigned to different types. During the Pleistocene cold periods, there were two other ice sheets in the northern hemisphere: the Laurentide Ice Sheet in North America and the Fennoscandian or Scandinavian Ice Sheet in northern and central Europe (► Chap. 8).

Ice caps are several orders of magnitude smaller than ice sheets (up to 50,000 km^2). Typical examples are the Devon Ice Cap in the Canadian Arctic or the ice cap on the North Island (Novaya Zemlya) in Russia.

Piedmont glaciers are formed when a glacier leaves the laterally constricting relief of a mountain range in order to then flow out in all directions in the foreland with little relief. Piedmont glaciers are not independent glaciers, but only parts of a larger glacier or glacier system. Again, as with outlet glaciers, delineation of upper glacier areas is problematic. Piedmont glaciers covered large areas of the Alpine foothills during the maxima of the Pleistocene glacial periods (e.g. Isar-Loisach glacier, Inn-Chiemsee glacier); the Malaspina glacier in Alaska is considered a prime example of a recent piedmont glacier.

> ❯ In the case of unconstricted glaciers, the glaciation is so strong that the subsurface no longer affects the direction of flow at the surface. Today, such situations only occur in the high latitudes, i.e. near or beyond the polar circles.

5.1.2.2 **Constrained Glaciers**

Constrained glaciers usually occur in high mountain ranges. They are restricted by the relief because their ice movement is controlled by the subglacial valley topography. The strongest conceivable type of glaciation in a mountain range is the **ice field (dendritic glacier system)**. This has already been discussed in the context of the nutritional glacier types; at this point not only the glacier types but also the classification systems overlap. The ice fields of the glacial periods fed the piedmont glaciers just mentioned; recent examples can be found in the Elias chain in Alaska or in Spitsbergen.

When the degree of glaciation decreases and the glaciers still reach more or less far down into valleys (◘ Fig. 5.4), but are no longer connected to those of neighbouring valleys via passes and watersheds, they are referred to as **valley glaciers**. These are the typical larger glaciers of the high mountains, which occur in all climatic zones. They are fed by one or more firn basins.

If glaciers do not have tongues but are confined to small flattenings called cirques (▶ Sect. 10.4), they are called **cirque glaciers** (◘ Fig. 5.5). These glaciers are the first to form when a mountain range glacierizes and the last to remain before it becomes ice-free. In this case, they arise from valley glaciers that lose their tongues due to area shrinkage. Cirque glaciers mark the climatic existence limit for glacierization.

◘ **Fig. 5.4**   Valley glaciers in Central Asia: Muskulak Glacier in the Pamirs (top) and Abramov Glacier in the southern Alay Range, Tajikistan (bottom). (Photo above: Hans-Dieter Schwartz, photo below: W. Hagg)

◘ **Fig. 5.5**   Cirque glacier (Vedretta del Lupo) in the Orobie Alps. (Photo: W. Hagg)

◘ Fig. 5.6 Hanging glacier in the Altay. (Photo: W. Hagg)

⟩ When the climatic snow line falls, the highest flattened areas, the so-called cirques, are the first to glacierize. If the glaciers grow beyond this, they are referred to as valley glaciers, and if these connect with each other across passes, an ice field is created.

If small glaciers do not exist in flat relief but on mountain flanks (◘ Fig. 5.6), they are referred to as **hanging glaciers**. In order to hold on in very steep terrain, they must be frozen to the bedrock. This is only possible when the ice is cold, so in non-polar areas they are restricted to very high elevations. Ablation in hanging glaciers often takes place mainly or exclusively via ice avalanches.

5.1.3 Thermal Glacier Types

The thermal classification is derived from the temperature profile of glacier ice (▶ Sect. 2.3). A glacier that is consistently composed of cold ice well below the pressure melting point is thus called a cold-based or **cold glacier**. Because it is frozen to bedrock, no basal sliding can take place, and ice movement occurs solely by internal deformation. This type of glacier is found in polar regions, where thin glaciers may consist entirely of cold ice due to low air temperatures. The uppermost layers of a cold glacier naturally also take on temperatures in the range of the melting point in summer.

A warm-based or **temperate glacier** consists of ice that is at the pressure melting point. Basal sliding can occur on the meltwater film at the glacier bed. Temperate glaciers are found in the mountains of the mid-latitudes and the tropics. Here, except for very high altitude regions, temperate ice usually forms. During winter, the uppermost layers of temperate glaciers can cool well below the melting point.

A glacier does not have to be completely assigned to a single thermal category. Glaciers that have both temperate and cold areas are called **polythermal glaciers**. This type of glacier is often found in subpolar regions. It does not necessarily behave as one might intuitively think, namely that the cold areas are found at the top and the warm areas at the lower glacier tongue. In fact, the exact opposite is often observed: Since snow and firn are poor conductors of heat, they protect against the penetration of winter cold in the accumulation area. In addition, the melting and refreezing processes here in summer result in a supply of latent energy. Although energy must be expended during melting, it is released again when the

5

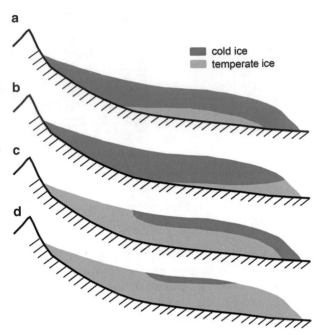

■ **Fig. 5.7** Different forms of polythermal glaciers. (**a**) mainly cold, temperate ice due to strain heating exists only near the glacier bed. (**b**) mainly cold, temperate ice exists at the glacier bed and at the snout. (**c**) mainly temperate, cold ice exists only near the surface in the ablation area. (**d**) mainly temperate, cold ice exists only near the surfeca in the upper part of the ablation area (Modified after Pettersson 2004)

cold ice
temperate ice

meltwater freezes in deeper snow layers. This leads to a warming of the snow pack, which is then close to the melting point. This temperate snow eventually becomes temperate firn and finally temperate ice. In the ablation area, although melting and refreezing also occurs in spring as long as a snow cover exists, thereafter this latent heat gain is absent. In addition, the thinner snow cover is a poorer insulator in winter, so that the glacier tongue can cool more in winter and therefore consists of cold ice.

The type of polythermal glacier just described is called the Svalbard type (■ Fig. 5.7c, d). However, there are also arrangements of cold and warm areas that deviate from this. In the Canadian High Arctic, for example, the ice is predominantly cold and small temperate zones exist only at the base (■ Fig. 5.7a, b).

5.2 **Distribution of Glaciers**

The global glacier ice reserves contain about 74% of the world's freshwater reserves and cover about 10% of the continents. For a long time, the number of glaciers on Earth was unknown or varied greatly depending on the study. This is due to the fact that in glacier inventories the minimum size up to which smaller ice patches are taken into account varies. In addition, the number of glaciers changes over time due to the disappearance of small glaciers, but also due to the break-up of larger glaciers into several smaller ones. Here, of course, it depends on the method of counting, i.e. whether the parts of glaciers that break up are still considered as one unit or as several glaciers. The global glacier area is also not easy to quantify. A first attempt, based on maps, aerial and satellite images around the 1960s, is the World Glacier Inventory (WGMS 1989). This project recorded 100,000 glaciers with a total area of 240,000 km^2, which, however, corresponds to only about one third of the total global area. Problems in remotely sensed mapping of glaciers are posed by clouds, snow and debris cover, which make delineation difficult and fully automatic mapping impossible. Even with semi-automatic methods, manual post-processing and correction are still necessary, which represents a considerable effort for global inventories.

A more recent project initiated at the beginning of this century is GLIMS (Global Land Ice Measurements from Space). More than 60 institutes are participating internationally in order to create a worldwide inventory through a standardised procedure on a homogeneous data basis. However, the first inventory with global coverage is the RGI (Randolph Glacier Inventory), which was prepared for the fifth assessment report of the Intergovernmental Panel on Climate Change (IPCC 2013). Here, the focus was on completeness of coverage rather than homogeneity of the database as in GLIMS. Currently, both inventories are being merged. According to version 6 of the RGI (2017), there are 215,547 glaciers worldwide with a total area of 705,739 km^2 (excluding the two ice sheets), the distribution of which is shown in (◘ Fig. 5.8).

The average glacier area in different regions varies considerably: in the Russian Arctic it is 48.2 km^2, in New Zealand only 0.33 km^2 (RGI 2017). Globally, in terms of glacier number, the maximum is in the 2–4 km^2 size class, but most of the area is contained in relatively large glaciers in the 256–512 km^2 class (Pfeffer et al. 2014). In the non-polar high mountains, the imbalance in glacier size is particularly pronounced (◘ Fig. 5.9).

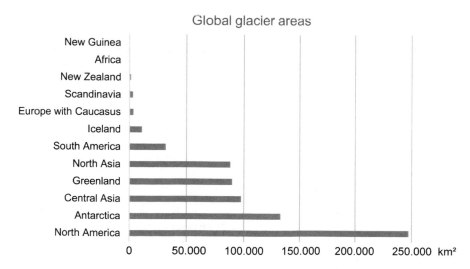

◘ **Fig. 5.8** Regional distribution of glaciers and ice caps (excluding ice sheets). (Data source: RGI 2017)

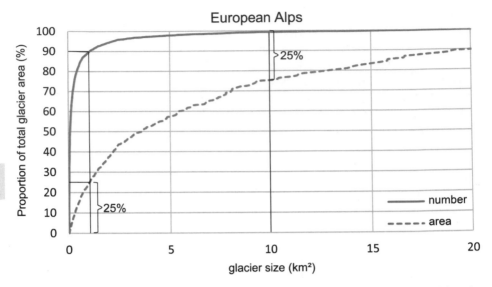

○ Fig. 5.9   Cumulative share of different glacier sizes in the total number and area of Alpine glaciers. (Data source: RGI 2017)

There are still 3896 glaciers in the Alps (RGI 2017). 3489 or 90% of them are smaller than 1 km², but contribute only 25% to the total area. In contrast, only 27 glaciers larger than 10 km² still exist in the Alps. Although this represents only 0.68% of the number, these glaciers also account for 25% of the total glacier area.

References

Ahlmann HW (1935) Contribution to the physics of glaciers. Geogr J 86:97–113

IPCC (2013) Climate change 2013: the physical science basis. Contribution of working group I to the fifth assessment report of the intergovernmental panel on climate change [Stocker TF, Qin D, Plattner G-K, Tignor M, Allen SK, Boschung J, Nauels A, Xia Y, Bex V, Midgley PM (eds)]. Cambridge University Press, Cambridge, UK/New York

Pettersson R (2004) Dynamics of the cold surface layer of polythermal Storglaciären, Sweden. Ph.D. thesis, The Department of Physical Geography and Quaternary Geology, Stockholm University

Pfeffer WT, Arendt AA, Bliss A, Bolch T, Cogley JG, Gardner AS, Hagen J-O, Hock R, Kaser G, Kienholz C, Miles ES, Moholdt G, Moelg N, Paul F, Radic V, Rastner P, Raup BH, Rich J, Sharp MJ, Andreassen LM, Bajracharya S, Barrand NE, Beedle MJ, Berthier E, Bhambri R, Brown I, Burgess DO, Burgess EW, Cawkwell F, Chinn T, Copland L, Cullen NJ, Davies B, De Angelis H, Fountain AG, Frey H, Giffen BA, Glasser NF, Gurney SD, Hagg W, Hall DK, Haritashya UK, Hartmann G, Herreid S, Howat I, Jiskoot H, Khromova TE, Klein A, Kohler J, Koenig M, Kriegel D, Kutuzov S, Lavrentiev I, Le Bris R, Li X, Manley WF, Mayer C, Menounos B, Mercer A, Mool P, Negrete A, Nosenko G, Nuth C, Osmonov A, Pettersson R, Racoviteanu A, Ranzi R, Sarikaya MA, Schneider C, Sigurdsson O, Sirguey P, Stokes CR, Wheate R, Wolken GJ, Wu LZ, Wyatt FR (2014) The Randolph glacier inventory: a globally complete inventory of glaciers. J Glaciol 60(221):537–552

RGI Consortium (2017) Randolph glacier inventory (RGI) – a dataset of global glacier outlines: version 6.0. Technical report, global land ice measurements from space, Boulder, Colorado, USA. Digital Media. https://doi.org/10.7265/N5-RGI-60

Schneider H-J (1962) Die Gletschertypen. Versuch im Sinne einer einheitlichen Terminologie. Geographisches Taschenbuch, Wiesbaden, pp S 276–S 283

Visser PC (1934) Benennung von Vergletscherungstypen. Z f Glkd 21:137–139

von Klebelsberg R (1948) Handbuch der Gletscherkunde und Glazialgeologie, vol 1. Springer, Wien

World Glacier Monitoring Service (WGMS) (1989) World glacier inventory: status 1988. In: Haeberli W, Bösch H, Scherler K, Østrem G, Wallén CC (eds) IAHS(ICSI)/UNEP/UNESCO. World Glacier Monitoring Service, Zürich

Glaciers and Climate

Contents

© The Author(s), under exclusive license to Springer-Verlag GmbH, DE, part of Springer Nature 2022
W. Hagg, *Glaciology and Glacial Geomorphology*,
https://doi.org/10.1007/978-3-662-64714-1_6

Overview

"Glaciers are melting because it's getting warmer". Although this sentence is not completely false, a more detailed exposition of the context makes it clear that it falls short, to say the least. Both "the climate" and its change and the response of glaciers can be quite complex. For one thing, air temperature is neither the only nor the most important climate element affecting glacier behavior. For another, the speed and magnitude of the response can both differ in different climates and be driven by non-climatic variables, making simple attribution difficult. Finally, glacier advances also exist without any climatic cause, which entails the risk of a complete misinterpretation.

6

Due to the current changes in both the cryosphere and the atmosphere, the term "glacier" is currently often mentioned in the same breath as terms such as "climate", "climate change" and "global warming". Before examining this connection in more detail, the term "climate" will be briefly discussed here. "Climate" is defined as the average state of the atmosphere in a particular place or area. To ensure that extreme values do not distort the statistics, a sufficiently long period of time must be taken as a basis. This is usually 30 years, the so-called climatological reference period. The World Meteorological Organization (WMO) has once defined the period 1961–1990 as the reference or normal period; in 2021 it has been replaced by the years 1991–2020. Strictly speaking, only 30-year periods may be compared when considering climate change. Here, not only mean values may change, but also other statistical quantities such as dispersion or extreme values. The term "weather", on the other hand, describes the state of the atmosphere at a point in time or a very short period of time at a specific location. Weather and climate are characterised by quantifiable atmospheric phenomena such as air temperature, precipitation, humidity, air pressure, wind direction and wind speed.

❯ "Climate" describes the average state of the atmosphere over a long-term period, while "weather" describes what is happening at a particular time.

6.1 Climatic Control of Glacier Behaviour

In ▶ Chap. 3 the climatic conditions for the formation of glaciers were discussed and in ▶ Chap. 4 it was shown how the altitude of the equilibrium line (ELA) separates the zones that are decisive for the mass balance. And it is precisely this elevation of the equilibrium line, which depends on meteorological conditions, that controls the mass balance and thus glacier behaviour. It is determined by the winter snow accumulation and the summer loss of snow, firn and ice, and can take on different values each year depending on the weather conditions. In a stable climate, it fluctuates around a stable mean value and controls the annual mass balance, which also fluctuates around a mean value (here: zero). However, if the climate changes, a process chain is set in motion that ultimately also changes the glacier.

In the event of glacier-favourable changes, i.e. an increase in snowfall or a decrease in melt rates in cooler summers, the snow line moves downwards. This means that the snow survives the summer down to lower elevations. On the glacier, the equilibrium line also moves downward, increasing the accumulation area at the expense of the ablation area, and increasing mass gains by accumulation. If this happens several years in a row, then the ice thickness in the accumulation area increases. The thickening leads to an increase in flow velocity and the excess mass is transported through the glacier. When it reaches the end of the glacier, more mass arrives there than melts off, and the tongue responds by advancing.

In the opposite case, as we have predominantly observed for several decades, the snow line moves upwards. Consequently, the entire process chain also reacts in the opposite way, and the glacier tongue reacts by melting back. Furthermore, it remains to be stated that only the elevation of the equilibrium line and the glacier mass balance are directly related to climate. They are the direct, unfiltered and undelayed response to meteorological conditions in the budget year. Changes in glacier length due to advance or retreat and associated changes in glacier area respond with a delay because the mass signal must first be transported throughout the glacier. The time delay from the climate signal to the response of the glacier tongue is called **terminus response time.** It is easy to understand that glaciers with fast mass turnover also react quickly to climate fluctuations, i.e. have short response times.

Rapid mass turnover is mainly associated with small and fast-flowing glaciers. Since the flow rate depends primarily on temperature regime, mass input and slope, it is mainly temperate, maritime and steep glaciers that have short response times. An example of this is the Franz Josef Glacier in New Zealand (❏ Fig. 6.1), which has an extremely short response time of only 3–4 years despite a considerable length of 10.5 km (Purdie et al. 2014).

❯ The climatic elements directly control the elevation of the equilibrium line and thus the mass balance. Changes cause the ice transport to strengthen or weaken until, after a reaction time whose duration depends on size and flow velocity, the glacier front responds.

❏ **Fig. 6.1** The tongue of Franz Josef Glacier in 2012; by 2019 it had receded by about 800 m from this position. Three people in the centre of the image illustrate the scale. (Photo: W. Hagg)

In contrast, it takes longer for large, cold, flat and continental glaciers until the climate signal is expressed in a change in length. Oerlemans (2007) calculated the terminus response on four glaciers in the Alps and Norway. At two small, steep glaciers it was about five years, at two valley glaciers with shallow and long glacier tongues it was about 35 years. Consequently, what happens at the tongue of these glaciers is the response to climate three and a half decades ago. Even longer than the terminus response time is the **dynamic/volume response time**. This is the time it would take for a glacier to adjust to a sudden climate change and, with a new size and geometry, to find a new equilibrium. Since climate changes are seldom erratic, but extend over decades or longer, the adjustment time can only be determined theoretically using assumptions about geometry and ice dynamics. Values vary from decades to millennia. Contrary to older ideas, adaptation time does not appear to necessarily increase with glacier size, but rather to be controlled by continentality and climate zone. A comparative study in seven regions concluded that adaptation time is about three decades in extremely maritime mountains, about five to eight decades in temperate mid-latitude continental climates, and centuries to millennia in the Arctic (Raper and Braithwaite 2009). According to a recent study, Alpine glaciers need an average of 50 years for their geometry to adjust to a new climate (Zekollari et al. 2020).

Changes in glacier length or area are therefore more difficult to interpret climatologically than mass balances, because they are delayed and are more strongly influenced by non-climatic factors. This underlines the importance of glacier mass balance measurements and especially of long-term observations for climate-relevant statements.

6.2 Glaciers as Climate Indicators

Glaciers are rightly regarded as excellent climate indicators, and there are several reasons for this. For one thing, they can make even small changes in climate parameters very clearly visible. For some valley glaciers, a change in annual temperature of a tenth of a degree Celsius can cause a change in length of several hundred metres. On the other hand, they do not react to short-term weather caprices, but filter out these events and show us the longer-term trend, which is what climate-related issues are all about. Furthermore, glaciers provide information from elevations and areas of the world from which there are hardly any measurements. Even in well-developed mountain ranges such as the Alps, there are still too few long-term measurements from high-altitude stations for many climate change issues, and the situation in other high mountain ranges of the world is even more limited in this respect. Here, glaciers can contribute to closing knowledge gaps. It is particularly helpful here that length and area information can be determined relatively quickly and easily using remote sensing methods. But they are not only important for current climate change: because their former extent is indicated by moraines in the landscape, they also contribute to the reconstruction of the paleoclimate, knowledge of which is enormously important in order to be able to correctly assess current climate change.

The climatic interpretation of glacier variations is not an easy undertaking, despite the apparently strong causal relationship. Problems can arise because glacier behaviour always represents the response to the overall climate (e.g. radiation, temperature, precipitation, humidity) and a quantitative assignment to individual parameters is not always easy. In ▶ Chap. 4 it became clear that air temperature has a much smaller contribution to melting than solar radiation. However, the two climate elements correlate with each other because summers with high radiation are usually also hot summers.

A glacier advance can, for example, be caused by snowy winters, by cooler summers, or by both. Here it is enormously helpful if not only the annual mass balance is known, but also the summer and winter balance. The winter balance can be determined by determining the water equivalent of the snow cover at the end of the accumulation period, the summer balance then results from the difference to the annual net balance. Only a comparison of the seasonal balances answers the question of whether the annual balance, and thus the glacier behaviour, is controlled by snowfall or melt. In the case of the Aalfotbreen in Norway, it can be clearly demonstrated in this way that the glacier advances in the early 1990s were not caused by cooler summers, but by snowier winters: While the summer balance does not change significantly from 1980 to 1995, the dashed trend line shows a sharp increase in the winter balance (◘ Fig. 6.2). This information could not be derived from the annual net balances.

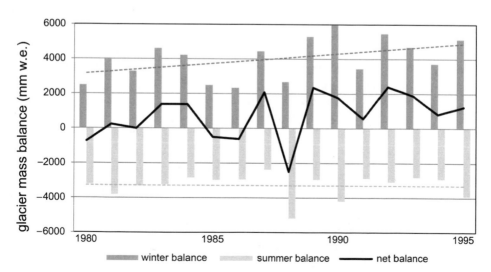

◘ **Fig. 6.2** Seasonal and annual mass balance at Aalfotbreen, Norway from 1980 to 1995. (Data source: WGMS 2019)

> The determination of the summer and winter balance has a strong additional benefit for climatological interpretations, because it makes visible which season is responsible for a specific glacier behaviour.

In the case of air temperature, it is not only the mean values that are decisive, but also the distribution: if high mean values result from individual very hot days, this can cause less mass loss than warm days over a longer period. It is also decisive whether precipitation falls on cold or warm days. As a result, snow amounts can differ significantly during two periods with the same mean values. But even in summer, individual precipitation events can have a decisive effect on the mass balance. Summer snowfalls abruptly set the ice melt to zero. The energy that has to be expended for renewed snowmelt does not go into ice melt and thus reduces the summer mass losses.

It must also be taken into account that the climate sensitivity of glaciers depends on the regional climate, the so-called macroclimate. In maritime climates with high precipitation, winter precipitation is decisive for glacier behaviour. In continental regions, i.e. those far from the sea, where snowfall is generally lower, radiation in the summer months is more decisive. In tropical regions, on the other hand, the water vapour content of the atmosphere is much more important because it affects radiation totals and the albedo of ice surfaces. In southern Asia, the strength of the monsoon must be taken into account. In addition, here, as on all tropical glaciers, accumulation and ablation occur simultaneously (▶ Chap. 4). This exemplary list is intended to show that glacier behaviour is influenced by climate everywhere on Earth, but not everywhere in the same way.

One problem with the climatic interpretation of glacial history via moraine deposits is that one does not get any information about minimum extents. **Moraines** only ever mark the peak of glacial extent, which was never reached again afterwards. In between there can be huge gaps where the glaciers were much smaller. Smaller glacial advances also remain undocumented if the moraines are destroyed again by younger, more extensive advances. Moreover, the strength of an advance depends not only on the intensity of a climatic deterioration, but also on its duration and on the initial position of the glacier tongue. In other words, brief but severe climate deteriorations can cause glaciers to advance just as far as more moderate but prolonged cold spells. Patzelt (1973), for example, assumes that the cooling around 1920, which left behind smaller moraines in the Alpine region, was similar in intensity to that at the height of the Little Ice Age around 1850. It cooling just lasted not long enough to allow the glaciers to advance as far as they had 70 years earlier.

Unfortunately, glacier behaviour is also influenced by non-climatic factors, which can lead to misinterpretations. The most important factor here is topography, which determines the slope of small glaciers and the area-height distribution of large glaciers. For shallow glaciers, a rise in the equilibrium line affects larger proportions of area than for a steep glacier with greater vertical extent (◼ Fig. 6.3a). Consequently, a more sensitive response is expected from the flatter

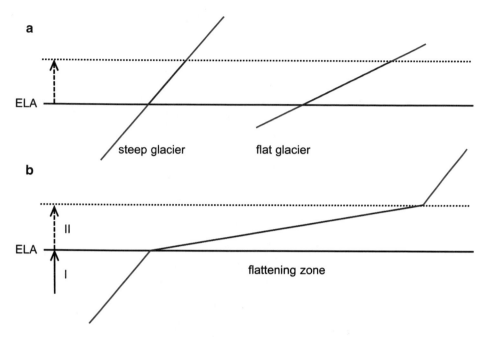

◘ Fig. 6.3 Influence of topography on glacier behaviour. (**a**) Flat glaciers react more sensitively because larger surface areas (red) are affected by an increase in the equilibrium line altitude (ELA). (**b**) A further increase in ELA (II) by the same amount as (I) would lead to greater mass losses than the previous increase

glacier. For large glaciers, the elevation of flattening zones is crucial. In their altitudinal range, the glacier reacts disproportionately to fluctuations of the equilibrium line (◘ Fig. 6.3b).

Due to the intensifying global warming, glaciers are now in a clear imbalance and lag far behind the climatic development. This is currently exacerbating the negative mass balances. Because the glaciers cannot melt as fast as the atmosphere warms, some glaciers are still located in very low and warm areas, where they cannot be explained by today's climate and would also disappear if there were no further warming. Such areas naturally have extremely high melt rates and make the mass balance more negative than it would be with an adjusted glacier size (Charalampidis et al. 2018).

Finally, there are also glacial advances that have no climatic cause and whose moraines can therefore be misinterpreted. These are, on the one hand, surges (▶ Chap. 3), in which a pent-up mass imbalance is suddenly relieved, and, on the other hand, advances triggered by mass movements. When rock or landslides fall on glaciers, this can lead to mechanical deformation of the ice due to the kinetic energy and pressure of the rock masses (▶ Excursus 6.1). However, even minor events can have an impact if they cover larger parts of the glacier tongue with debris. The debris has an insulating effect and efficiently reduces the melt (▶ Sect. 7.1), so that a positive mass balance and subsequently an advance of the glacier tongue can result.

In summary, it can be stated that glaciers are good and important climate indicators, but that the climatic interpretation of glacier behaviour is not possible without background knowledge and should be reserved for experts.

> Not suitable as climate indicators are surging, heavily debris-covered or calving glaciers. In the latter case, marine factors such as currents, water temperature or water level have a strong effect on calving processes and partially decouple these floating glacier tongues from the regional climate.

Excursus 6.1: Waiho Loop Moraine

A prominent example is found in the New Zealand Southern Alps (◘ Fig. 6.4), where the Waiho Loop moraine of Franz Josef Glacier was long interpreted as a late-glacial Egesen moraine (▶ Chap. 8) and was considered evidence for climate synchrony between the northern and southern hemispheres (Denton and Hendy 1994). However, more recent sediment analyses (Tovar et al. 2008) and mass balance modeling (Reznichenko et al. 2011) suggested that a rock fall onto the glacier tongue reduced ablation enough to cause this significant advance.

◘ **Fig. 6.4**   The Waiho Loop moraine of Franz Josef Glacier is thought to have been formed by an advance triggered by a rockfall. (Google Earth)

References

Charalampidis C, Fischer A, Kuhn M, Lambrecht A, Mayer C, Thomaidis K, Weber M (2018) Mass-budget anomalies and geometry signals of three Austrian glaciers. Front Earth Sci 6:218. https://doi.org/10.3389/feart.2018.00218

Denton GH, Hendy CH (1994) Younger dryas age advance of Franz Josef glacier in the southern Alps of New Zealand. Science 264(5164):1434–1437. https://doi.org/10.1126/science.264.5164.1434

Oerlemans J (2007) Estimating response times of Vadret da Morteratsch, Vadret da Palü, Brikdalsbreen from their length records. J Glaciol 53(182):357–362

Patzelt G (1973) Die neuzeitlichen Gletscherschwankungen in der Venedigergruppe (HoheTauern, Ostalpen). Z Gletscherk Glazialgeol 9:5–57

Purdie H, Anderson B, Chinn T, Owens I, Mackintosh A, Lawson W (2014) Franz Josef and fox glaciers, New Zealand: historic length records. Glob Planet Chang 121:41–55. https://doi.org/10.1016/j.gloplacha.2014.06.008

Raper SCB, Braithwaite R (2009) Glacier volume response time and its links to climate and topography based on a conceptual model of glacier hypsometry. Cryosphere 3:183–194. www.the-cryosphere.net/3/183/2009/

Reznichenko NV, Davies TR, Alexander DJ (2011) Effects of rock avalanches on glacier behaviour and moraine formation. Geomorphology 132:327–338

Tovar DS, Shulmeister J, Davies TRH (2008) Evidence for a landslide origin of New Zealand's Waiho loop moraine. Nat Geosci 1:524–526

WGMS (2019) Fluctuations of glaciers database. World Glacier Monitoring Service, Zurich. https://doi.org/10.5904/wgms-fog-2019-12

Zekollari H, Huss M, Farinotti D (2020) On the imbalance and response time of glaciers in the European Alps. Geophys Res Lett 47:e2019GL085578. https://doi.org/10.1029/2019GL085578

Glaciers and Water

Contents

© The Author(s), under exclusive license to Springer-Verlag
GmbH, DE, part of Springer Nature 2022
W. Hagg, *Glaciology and Glacial Geomorphology*,
https://doi.org/10.1007/978-3-662-64714-1_7

Overview

Glaciers consist of frozen water. Most ablation processes are accompanied by a transition to the liquid phase, and to a lesser extent also to the gaseous phase. Hydrological systems exist on, in and under glaciers that regulate the budget of this water reservoir. In the case of supraglacial ice melt, in addition to the energy exchange with the atmosphere, the debris layers are of particular importance because, depending on their thickness, they can promote or slow down the melt. In the intraglacial system, the seasonal development of drainage channels is important because it affects the rate of drainage and thus the whole hydrological behaviour. The occurrence of subglacial water is important mainly for the mode of movement of the ice and for the shaping of the bedrock. The total runoff from glaciers has special characteristics that affect the streamflow of glacier-fed rivers.

7

Glaciers are important water reservoirs; 69% of global freshwater reserves are currently in the form of glacial ice (Liu et al. 2011). In many regions of the world, they are not only suppliers of drinking water, but also have economic importance, for example for agricultural irrigation or for the generation of hydropower. However, water in glacier environments can also become a natural hazard (▶ Chap. 9); in these cases, it therefore becomes the focus of science and disaster risk management.

Melting is the most important ablation process on many glaciers, which is why meltwater is ubiquitous on glaciers, at least in summer. Melting processes occur almost exclusively on the ice surface (**supraglacial**), but to a lesser extent they can also occur within the glacier (**englacial**) and on its underside (**subglacial**). The presence of meltwater at the base of the glacier, in turn, controls ice dynamics (▶ Chap. 3) and affects the shaping of the bedrock (▶ Chap. 10). An overview of the hydrological systems on and in a glacier is provided by (**◘** Fig. 7.1).

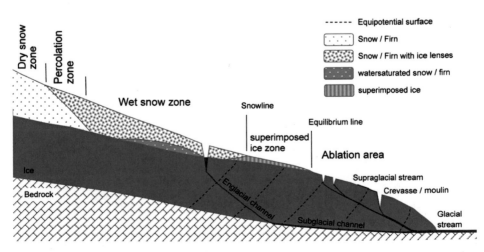

◘ Fig. 7.1 Overview of glacial hydrological systems. (Modified after Müller (1962), Röthlisberger and Lang (1987), Paterson (1994))

Glacial hydrology deals on the one hand with qualitative and quantitative phenomena of liquid water on and in glaciers, and on the other hand with the water balance of glaciated river basins. In this chapter, the different glacial hydrological systems are first described before the total runoff from glaciers is characterized and the hydrological importance of glaciers for river systems is explained.

7.1 Glacial Hydrological Systems

7.1.1 Supraglacial System

By far the largest part of the total melt takes place on the glacier surface, and different hydrological zones exist here with varying importance for the water balance (◘ Fig. 7.1).

Above the dry snow line is the **dry snow zone**, in which the air temperature never rises above 0 °C and therefore no melting takes place. Ablation in this zone is limited to sublimation and wind drift.

At the **percolation zone** below, melting takes place at the snow surface. When gravity is greater than the binding forces in the pores, meltwater seeps into deeper snow layers, which is called percolation. There it refreezes, resulting in the formation of ice lenses and the transport of latent heat from the surface downward. Refreezing occurs before the meltwater has percolated through the entire winter snowpack. Infiltration and refreezing are important processes by which meltwater can be temporarily stored in snow.

From the wet snow line onwards, the entire snow pack down to the autumn surface of the previous year is affected by melting and refreezing processes. In this **wet snow zone,** the entire snow layer is at melting temperature. On subpolar glaciers, the so-called **slush zone** often develops here, an extremely water-saturated layer above water-impermeable, cold ice. Here, sudden runoff events, so-called slush flows, can occur. In the lower part of the wet snow zone, the meltwater content is so high that it accumulates on the previous year's horizon and refreezes as compact, transparent ice (superimposed ice).

If this ice comes to the surface through the melting of the overlying snow cover, it forms the **superimposed ice zone**. This occurs to a significant extent only on subpolar glaciers. Here it is bounded above by the snow line and below by the equilibrium line. These two lines, which are normally identical, diverge here. The superimposed ice zone still belongs to the accumulation area, but is no longer snow-covered. This circumstance must be taken into account in mass balance measurements and makes it difficult to locate the equilibrium line, which here is marked not by the boundary between snow and ice, but between clear superimposed ice and granular glacier ice.

Below the equilibrium line is the **ablation area**. This is snow-free in summer, and due to the lower albedo of ice, particularly high melt rates are achieved here. The meltwater cannot be stored here, but flows off immediately in a network of small and larger supraglacial streams (◘ Fig. 7.2).

◘ **Fig. 7.2** Meandering meltwater stream on the tongue of Golubin Glacier, Kyrgyz Alatau. (Photo: W. Hagg)

7

❯ Snow and firn act hydrologically like a sponge: they can absorb liquid water and store it temporarily, which dampens runoff peaks.

The ablation area may consist of bare glacier ice on the surface or be covered by rock particles of various sizes, from dust to metre-sized boulders. When little or no ice is visible, one speaks of a "debris-covered glacier". Due to the emergent ice movement in the ablation area, which is directed against the surface (▶ Chap. 3), any rock that has reached the ice surface as a result of this movement or through rockfall remains on the surface until it is finally deposited at the glacier margin. In addition, dust may be blown onto the glacier by wind. If the aggregation of rocks and particles due to melt-out, rockfall and aeolian sedimentation at a certain location is greater than the removal by ice movement, debris layers or **supraglacial moraines** form, which have different hydrological effects depending on their thickness. Due to the decreasing ice movement during the current glacier retreat, debris-covered glaciers are currently forming in many areas of the world or existing debris layers are increasing in extent and thickness (Mayr and Hagg 2019).

Melt beneath a layer of debris is called **sub-debris ablation**. Compared to uncovered ice, a thin layer of debris enhances melting because debris absorbs more solar radiation due to its lower albedo, and it can also heat up more than ice. However, as the overburden gets thicker, less and less heat can be conducted through it. Therefore, at a certain thickness, known as the "critical debris thickness", the melting returns to that of bare ice. Depending on the properties of the debris (grain size composition, porosity, water content, etc.), the critical thickness lies between approx. 2 cm and 8 cm. With even thicker debris, the insulating effect predominates and ice melting is reduced compared to that of clean ice. These relationships can be visualized in the Østrem curve named after its first describer (◘ Fig. 7.3) (Østrem 1959).

❯ Ice melt is enhanced by thin debris layers and reduced by thick ones.

The aeolian sediment, i.e. the fine dust blown in by wind, is called **cryoconite** on glaciers and is glued together by microorganisms such as bacteria or blue-green

◘ Fig. 7.3 Change in melting rate under debris layers of different thicknesses

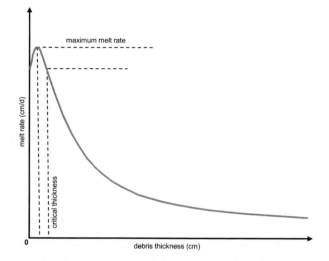

◘ Fig. 7.4 Cryoconite holes. The close-up on the top right shows the granular structure of the cryoconite. (Photos: W. Hagg)

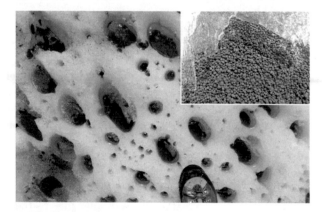

algae to form cryoconite granules. These granules can accumulate locally as a result of melt runoff. They also, if they form only a thin overlay, reinforce the melt. This creates cylindrical pits called **cryoconite holes** (◘ Fig. 7.4). These microforms of selective ablation can grow in depth only as long as solar rays strike the bottom of the hole. They therefore have a maximum depth that depends on the maximum solar altitude. On tropical glaciers they therefore grow deeper than on polar ones.

Melting holes also form from individual stones up to a maximum diameter. Above this diameter, the additional heating can no longer be conducted through the stone; thus, also larger stones or boulders develop an insulating effect. As so-called **glacial tables,** they grow out of their surroundings, where ablation is greater than below them (◘ Fig. 7.5). They also have a maximum height, because at some point the ice pedestal is so high that the sun's rays reach it and it begins to tilt. At some point the stone slips off the pedestal, which then quickly disappears, while the stone on the new base begins again to rise above its faster-melting surroundings.

Fig. 7.5   Glacier table in the northern Tienshan. (Photo: W. Hagg)

Debris cover, where present, generally increases towards the front of the glacier. Heavily debris-covered glaciers can have debris layers of 3 m thickness at their lower end; on the Miage glacier in the Italian Aosta Valley, trees even grow on the supraglacial moraine. The ice under such thick debris covers usually moves little and the glacier front is often very stable; mass losses here usually manifest themselves in a lowering of the surface rather than in length changes. As a final consequence, these heavily debris-covered glacier ends can separate from the active glacier and continue to exist as dead ice for a relatively long time.

Opinions in the literature differ on the efficiency of melting under thick debris layers. The fact that ablation still takes place at all is often due to the action of supraglacial meltwater streams, which regularly change their course and on whose banks so-called **ice cliffs** are formed. These are ice sections that are so steep that the entire debris layer slides off, leaving at most a thin layer of dirt behind. Together with supraglacial lakes, the formation of which is also related to meltwater streams and on the shores of which steep walls without debris layers also develop (■ Fig. 7.6), ice cliffs are considered "hot spots" of ablation on debris-covered glaciers (Buri and Pellicciotti 2018).

These meltwater streams, which are obviously important for the formation of ice cliffs, do not necessarily run supraglacially to the end of the glacier, but often disappear at vertical potholes, the **glacier mills** or **moulins**, inside the ice (■ Fig. 7.7).

Moulins preferentially form at crevasses and may persist after the crevasse has closed again. Mills can run dry if the stream is tapped by a new mill further upstream.

7.1.2 **Intraglacial System**

Intra – or englacial drainage exists only in temperate ice, while cold ice is impermeable to water and therefore only supraglacial drainage occurs on cold glaciers.

In temperate glaciers, however, fine cracks and fissures form between the crystals. Along these, low permeabilities develop in the intact glacier ice, which is called primary permeability. The thermal energy of liquid water causes preferred flow paths to widen more and more. When the water eventually flows in passages

◨ Fig. 7.6 Supraglacial lake
with ice cliffs on Abramov
Glacier, Kyrgyzstan. The
boundary between vertical
wall and inclined cliff marks
a former lake level. (Photo:
W. Hagg)

◨ Fig. 7.7 A meltwater
stream disappears into the
ice. Glacier mill on the
Pasterze, Austria's largest
glacier. (Photo: W. Hagg)

and tunnels, this is called secondary permeability. Eventually, a system of englacial conduits is formed, so meltwater drainage can be compared to drainage in a karst aquifer. A karst aquifer is a body of groundwater in soluble rock in which a system of underground tubes, caves and tunnels also develops. A possible division into a vadose zone, where the tubes are not completely filled with water and atmospheric pressure prevails, and a phreatic zone below the water table, where water flows under hydrostatic pressure, emphasizes the analogy to the karst water body (Sugden and John 1976). Opposing forces are always at work in the channels: ice creep from the outside acts to close the channels, while melting from the inside acts to widen them. The diameter of such tubes, which are also called **Röthlisberger-** or **R-channels**, is normally approximately circular (Shreve 1985); under various conditions, however, it can also deviate from this shape (❑ Fig. 7.8).

During the winter, the overburden pressure prevails and the tunnel system is destroyed; in the next melting period, it forms anew. In summer, the cross-sections of the tubes are efficiently widened by mechanical erosion and frictional heat, so that drainage becomes more efficient towards the end of the ablation period.

7

❑ **Fig. 7.8** Drained, englacial drainage pipe (Röthlisberger channel), already elliptically deformed by ice creep. Vernagtferner, Austria. (Photo: W. Hagg)

Specifically, this means that meltwater stays in the glacier for less time and the daily delay between maximum melt and maximum discharge at the glacier gate becomes shorter.

> Drainage channels that form in the ice and expand over the melt period are closed again in the absence of water pressure in winter and must reform in spring. For this reason, the speed and efficiency of water drainage have a seasonal cycle with a maximum in midsummer.

Intraglacial drainage in water-filled channels resembles a pressure flow from high to low hydraulic potential. This potential depends on the elevation (potential energy), the overburden pressure and the channel cross-section. The planes connecting points with the same hydraulic potential (equipotential surfaces) therefore do not run parallel to the glacier surface, but rise downglacier (◘ Fig. 7.1). The channels follow the steepest hydraulic gradients at right angles to the equipotential surfaces and can also flow uphill. As a rule, however, the drainage is more or less directed downwards, so that the channels eventually reach the glacier bed.

7.1.3 Subglacial System

The type of subglacial drainage is of crucial importance for the flow velocity and for geomorphological processes (► Chap. 11). Depending on the amount of water, basal temperature, topography and permeability of the subsurface, different drainage types can form. Generally, a distinction is made between discrete drainage systems, where water flow is linearly constrained to channels and distributed drainage systems, which enable are more diffuse or extensive flow. The study of drainage under glaciers is complicated by the fact that the processes involved usually elude direct observation. For this reason, some assumptions are based on theoretical considerations. An additional complication is that several types can be formed simultaneously on one glacier. (Annual) temporal changes in the systems are also possible (Bennet and Glasser 2009).

The normal case in temperate glaciers is a dentritic subglacial channel network. Water can flow rapidly in this discrete system, so that the glacier is rapidly drained. Channels formed only in the ice are again called Röthlisberger or **R-channels**, whereas those cut into the glacier bed are called **Nye-** or **N-channels**.

Distributed drainage systems are much more ineffective than channel systems and they can drain in a variety of ways. The water may be located in unconsolidated subglacial sediments, where it is moved by the deformation of the glacier bed, by free percolation in the pore aquifer, or along pipes. On bedrock, subglacial runoff may occur as a millimetre-thin film of meltwater between ice and subsurface (Weertman 1972). This water does not usually originate from supraglacial or englacial melt, but is subglacial in origin. Where energy gain from geothermal heat flow and friction is greater than energy dissipation, basal melting can occur. However, most of the meltwater film is formed by pressure melting at the stoss side of rock obstructions (► Sect. 3.2.2).

> Runoff under glaciers is either concentrated in channels or extensive. In the latter case, significantly less water flows because it only forms a thin film between ice and rock or seeps very slowly through the loose bedrock.

Transitional forms between discrete and distributed drainage systems are braided canal systems (Walder and Fowler 1994) or linked cavity systems (Kamb 1987). Braided canal systems develop from linear canal systems where the underground is soft and water is forced laterally into the contact zone between the ice and the glacier bed, creating wide, shallow, interconnected canals. Linked cavity systems consist of cavities connected by constrictions. Depending on water pressure, this network pattern is subject to constant change as connections close and new ones open, allowing individual cavities to be connected to and disconnected from the drainage system (Kamb 1987). At high discharges, water can become increasingly concentrated in a few flow paths, resulting in a dendritic Röthlisberger channel system.

7.2 Runoff from Glaciers

In hydrology, the term "runoff" refers to the discharge of water at a specific point in a stream or river course. The entire area from which water flows to this point is called the catchment area and is delimited from neighbouring catchment areas by the so-called watershed. The ice-covered portion of a catchment is called **glacierization**. Looking at large river basins, glacierization is usually low. In that part of the Alps drained by the Danube, glaciers exist with an area of 358 km² (Weber et al. 2016); for the entire catchment of the river at its mouth into the Black Sea, this corresponds to a glacierization of only 0.04%. The quantitatively most important alpine Danube tributary is the Inn. When it enters the Danube in Passau, it dominates the river characteristics. Its greater abundance of suspended matter **(glacial milk)** and its lower temperature compared to the Danube impressively prove its origin in the glacierized Central Alps in summer, although glacierization at the Wasserburg gauge is only about 5%. With increasing proximity to the glaciers, the area share of the glaciers and thus the glacial characteristics of the discharge steadily increase. The most extensive glacier areas (approx. 130 km²) in the Inn river basin are located in the Ötztal Alps; the glaciation of the Rofenache is already approx. 35% at the Vent gauge. Here, one third each of the water comes from ice melt, snow melt and rain (Weber et al. 2010).

  The highest discharge gauging station in the Ötztal and in the entire Danube catchment is the Vernagtbach gauging station at 2640 m above sea level, where 60% of the catchment area (as of 2020) is covered by glacial ice. This high proportion of area causes a runoff pattern strongly influenced by ice melt, which is also referred to as a glacial runoff regime. Typical for this is a clear maximum of the annual course during the strongest melt rates in July or August, while only a very low base flow occurs in the winter months (☐ Fig. 7.9a). This is due not only to the greatest supply of energy in the form of short-wave radiation and sensible heat at this time of year, but also to the minimum snow cover. At the same rate as the snow line

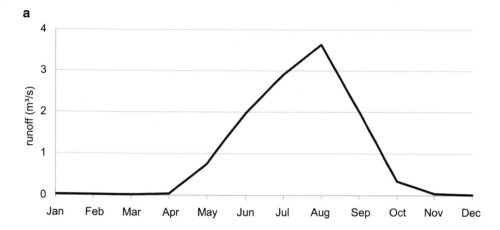

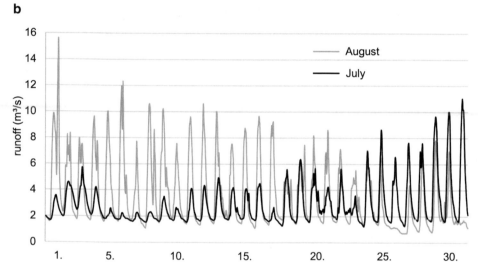

☐ **Fig. 7.9** Measured discharge at the Vernagtbach gauging station (glacierization: 60%) in 2018. Monthly mean values (**a**) and hourly values (**b**) of the months July (black) and August (grey). (Data source: Geodesy and Glaciology, Bavarian Academy of Sciences and Humanities)

moves upward over the summer, the darker ice surfaces increase in size. Because of the lower albedo of ice, it can absorb more radiation, so a more snow-free glacier in late summer melts much more than in early in spring, even if days with comparable solar radiation and air temperatures are compared. The albedo effect is impressive during summer snowfalls, which abruptly set the ice melt to zero and thus drastically reduce the total runoff.

In addition to the strong seasonal fluctuations, high discharge amplitudes during a diurnal cycle are typical for glacial runoff. The maximum discharge occurs, with a certain delay to the maximum melt, in the afternoon, while the lowest water levels in the glacial stream are recorded in the early morning. This has been the undoing of many a hiker who has crossed such a stream early in the day and made his way back in the afternoon. The diurnal variations at Vernagtferner become more pronounced over July (■ Fig. 7.9b), because the ablation area becomes larger as the snow line moves upwards and the steadily developing englacial channel system drains more effectively. At the end of July, the discharge varies in the diurnal cycle by a factor of five, the maximum on an afternoon in early August is almost eight times as high as the minimum on the same morning.

❯ Glacial streams have both a very strong diurnal cycle between morning minimum and afternoon maximum and a very strong annual cycle between winter minimum and summer maximum.

As far as the year-to-year variation of runoff is concerned, glaciers have a balancing effect on water flow. In cool and wet years, they build up their reserves, which they use up again in dry and hot years and add to the runoff. Due to the fact that the glaciers melt heavily just when there is no precipitation, rivers with glaciers in the catchment area always have a guaranteed minimum amount of water, which is generated either by rain or by melting. This **compensation effect** (Röthlisberger and Lang 1987) is most pronounced for glaciations around 40%. In the case of very strong or very low glaciations, one of the two components (rain or melt) dominates, which means that the runoff volumes are again subject to greater interannual fluctuations.

❯ Further downstream, glaciers have a regulating effect on river levels, because they melt especially in hot, dry summer periods when rain is lacking. Glacier-fed rivers therefore have smaller fluctuations in discharge from year to year and are less threatened by low water.

References

Bennet MR, Glasser NF (2009) Glacial geology. Ice sheets and landforms. Wiley-Blackwell, Chichester

Buri P, Pellicciotti F (2018) Aspect controls the survival of ice cliffs on debris-covered glaciers. Proc Natl Acad Sci:201713892. https://doi.org/10.1073/PNAS.1713892115

Kamb B (1987) Glacier surge mechanism based on linked cavity configuration of the basal water conduit system. J Geophys Res 92:9083–9100

Liu J, Dorjderem A, Fu J, Lei X, Lui H, Macer D, Qiao Q, Sun A, Tachiyama K, Yu L, Zheng Y (2011) Water ethics and water resource management. Ethics and climate change in asia and the pacific (ECCAP) project, Working group 14 report. UNESCO, Bangkok

Mayr E, Hagg W (2019) Debris-covered glaciers. In: Heckmann T, Morche D (eds) Geomorphology of proglacial systems, geography of the physical environment. Springer, Cham, pp 59–71. https://doi.org/10.1007/978-3-319-94184-4_4

Müller F (1962) Zonation in the accumulation area of the glaciers of Axel Heiberg Island, N.W.T., Canada. J Glaciol 4(33):302–311. https://doi.org/10.3189/S0022143000027623

Østrem G (1959) Ice melting under a thin layer of moraine, and the existence of ice cores in moraine ridges. Geogr Ann 41:228–230

Paterson WSB (1994) The physics of glaciers, vol 481. Butterworth Heinemann, Oxford/Burlington

Röthlisberger H, Lang H (1987) Glacial Hydrology. In: Gurnell AM, Clark MJ (eds) Glacio-fluvial sediment transfer. John Wiley, Chichester, pp 207–284

Shreve (1985) Esker characteristics in terms of glacier physics, Katahdin esker system, Maine. Geol Soc Am Bull 96:639–646

Sugden DE, John BS (1976) Glaciers and landscape. Halsted Press, London

Walder F (1994) Channelized subglacial drainage over a deformable bed. J Glaciol 40:3–15

Weber M, Braun L, Mauser W, Prasch M (2010) Contribution of rain, snow- and icemelt in the upper Danube discharge today and in the future. Geogr Fis Din Quat 33:221–230

Weber M, Prasch M, Kuhn M, Lambrecht A, Hagg W (2016) Ice reservoir. In: Mauser W, Prasch M (eds) Regional assessment of global change impacts. The project GLOWA-Danube. Springer, Cham/Heidelberg/New York/Dordrecht/London, pp S 109–S 115. https://doi.org/10.1007/978-3-319-16751-0

Weertman J (1972) General theory of water flow at the base of a glacier or ice sheet. Rev Geophys 10:287–333

Glacial History

Contents

Overview
There is a whole range of methods that can be used to reconstruct former glacier length. Many of them are aimed at the dating of moraines, but bogs and human traces also allow conclusions to be drawn about past extents. In the ice ages, warm periods (interglacials) alternated with cold periods (glacials) during which the glaciers were particularly large. The triggering and regulating mechanisms of these large climate fluctuations are subject to both terrestrial and extraterrestrial control. The most recent ice age, the Pleistocene, has left visible traces in the landforms of Europe and North America to this day. In the period since the last glacial there have also been fluctuations of climate and glaciers, but with a smaller amplitude than in the epoch before. At present, we are dealing with accelerated glacier retreat almost everywhere on Earth, which will continue in the future, and the consequences of which act on different spatial scales.

8

Glacial history is the temporal development of glaciers in the past. Global glacier extent has always been subject to large fluctuations, like the global climate as the main driver. For the older past, before glacier behaviour could be directly observed by humans, former glacier extents have to be reconstructed by indirect methods. At the beginning of this chapter, a brief outline of these methods is given before the glacial history is highlighted on different time scales. The first scale concerns glaciation phases that occurred very far back in time and that no longer have any influence on the present shape of the Earth's surface. The second scale is the Pleistocene, i.e. the most recent ice age, which shaped the landscape of large areas of Europe and North America. The third scale is the glacial history of the Holocene, the period since the last ice age. Particular attention is paid here to modern fluctuations and current glacier retreat. Naturally, the younger the period under consideration, the better and more numerous the information on it.

8.1   Methods for the Reconstruction of Glacial History

During advances, glaciers pile up loose rock at their front and deposit debris that they carry with them on their sides (▶ Chap. 11). The resulting ice-marginal **moraines** provide direct evidence of former glacial extensions. If younger advances extend farther than older ones, then the old moraines will be destroyed as they pass over them. For this reason, not all advances are usually marked, and a glacial history based on moraines is usually incomplete. On the other hand, this circumstance already allows a relative dating when several moraine ridges are present: the outermost moraines are the oldest, with increasing proximity to the glacier the deposits become younger. Simple age determinations and correlations across valleys are possible by morphological criteria such as freshness of form and state of preservation. Steep moraines with sharp ridges are younger than strongly rounded and flattened forms.

❯ Moraines become younger and younger towards the glacier.

Further age indications are possible via stages of the soil-forming processes (e.g. humification, oxidation, acidification). The weathering grades of rocks also allows conclusions to be drawn about the age of the moraine. With the aid of a rebound hammer ("Schmidt hammer"), an instrument developed for materials testing, the hardness of the rock surface can be tested (◘ Fig. 8.1). Since this depends on how long the surface was exposed to atmospheric weathering processes, this method allows relative age dating (Winkler 2000, 2005). A relative age allows statements about which deposit is younger or older than another. Sometimes fossil soils buried by moraines are found. If these contain organic matter, this is a particularly lucky case, because then **radiocarbon dating** (▶ Excursus 8.1) can be used to determine the absolute age of the soil (in years before today) and thus the time of overburden.

Another possibility for dating moraines is offered by **lichenometry**. It is based on the constant growth of some lichen species, which allows to infer the age of the lichen from its diameter (Beschel 1950). Since exposed or deposited rock surfaces are colonized by lichens very rapidly, the largest diameters of lichens provide good indications of the depositional age of larger rocks and boulders. Since the rate of growth depends on macroclimate and rock, regional calibration curves must be

◘ **Fig. 8.1** Methods for dating moraines. Left: Application of the rebound hammer on a moraine in the Silvretta group. Right: The lichen *Rhizocarpon subgenus* (green-yellowish) next to other species (grey, reddish) on a boulder. (Photo: W. Hagg)

considered. These are constructed using lichen diameters on surfaces of known age (Locke et al. 1979). Structures or moraines that have been dated by other methods may be considered for this purpose. The applicability of the method is limited to a few centuries. Because of their longevity and constant growth rate, lichens of the genus *Rhizocarpon* (◘ Fig. 8.1) are particularly well suited for lichenometric studies (Armstrong 2016).

Fossil wood, i.e. tree remains that died long ago, has also made an important contribution to the reconstruction of glacial history. So-called in situ finds, which have grown at the site of discovery, are particularly valuable (Furrer and Holzhauser 1984). This is possible, for example, in the protection of rocks which prevent the tree from being carried away by the advancing glacier. In situ finds can be recognized by the fact that the rootstock is still well anchored in the ground, that the trunk still has numerous branches or that the bark is only scraped off on one side. Tree trunks that have been transported in glaciers usually have neither branches nor bark. The in-situ location shows the glacier extent at the time of death, the age of the tree proves the minimum period in which the glacier extent was smaller before. To this period can be added the average minimum time period between the disappearance of ice and recolonization. Depending on sea level and tree species, this varies between 5 and 10 years or 60 and 100 years (Holzhauser 2009).

If several in-situ woods are found, even the rate of advance can be calculated. While the age of the tree can be determined very simply by counting the annual rings, the time of death of old finds can be determined by radiocarbon dating of the outermost annual rings. However, it is much cheaper and more accurate to determine the age of a tree by means of tree ring research (dendrochronology). Here, both the width and the density of the individual rings are examined. By comparing the ring patterns of many trees, an average tree ring sequence (tree ring chronology) can be established. Due to overlaps of different tree ages, the sequence can amount to several thousand years. Living trees, timber used in buildings and fossil finds are used for this purpose. The result is a pattern typical for the region, reminiscent of a barcode. In the altitudinal range of the alpine timberline, where borderline living conditions for tree growth prevail, the tree ring chronologies even apply to the entire alpine region (Holzhauser 2009). In the case of a new find, it is possible to check which sequence of the barcode it represents, and the life phase of the tree can thus be dated to the year. In this way, it was possible to reconstruct the advance and retreat phases of the last 3500 years at the Great Aletsch Glacier (Holzhauser et al. 2005).

Landscapes that have become ice-free tend to form lakes. This can be explained by the fact that glacier-formed relief has many depressions and hollow forms (► Chaps. 10 and 11), the subsoil of which is relatively impermeable to water due to the fine material ground up by the glacier ("boulder clay"). Natural siltation processes result in the formation of bogs from lakes. These consist of incompletely decomposed plant remains (peat), which can be dated using the radiocarbon method and ensure a high temporal resolution due to the relatively rapid vertical growth. Dating the lowest layers gives the age of the bog and thus the minimum time period for the site to become ice-free. In addition, alluvial deposits of sand may indicate renewed glacial advances that have not reached the bog.

Fig. 8.2 Watercolour of the Rhone Glacier by Spengler (1843–1908). The picture probably dates from the 1890s (Heinz J. Zumbühl, personal communication); the bulging tongue suggests an advance. Today, the end of the glacier is no longer visible from the painter's point of view

Historical glacier fluctuations can also be dated by human traces. Destroyed paths, water intakes and buildings can give an indication of former glacier extensions; here, terrain archaeology can provide valuable clues. Written and pictorial sources, in collaboration with historians and art historians, can also add to the knowledge of glacier fluctuations. Written sources include chronicles, travelogues, and records of devastation. Pictorial sources include drawings and paintings (■ Fig. 8.2) as well as maps or, more recently, photographs (Zumbühl and Nussbaumer 2018).

In order to quantify and compare the strength of individual glacier advances and the associated climate deterioration, the distance of the glacier tongue or its terminal moraine (▶ Chap. 10) from the centre of the ice is simply used for ice sheets and ice caps (▶ Chap. 5). This is permissible because these glaciers flow more or less undisturbed over flat relief. Mountain glaciers, on the other hand, are strongly controlled by bedrock topography, which makes their length a poor indicator. In order to compare their advances with each other, the lowering of the equilibrium line (▶ Excursus 8.2), which was necessary to allow the glacier to grow to the corresponding extent, is used instead.

Excursus 8.1 Radiocarbon Dating

Of the three isotopes of carbon existing in nature, ^{12}C, ^{13}C and ^{14}C, the latter is formed in the upper atmospheric layers by cosmic ray bombardment of ^{14}N. ^{14}C, unlike the other carbon isotopes, is therefore radioactive and decays back to nitrogen with a half-life of 5730 years. Living organisms incorporate the carbon isotopes into their biomass in approximately the same ratio in which they occur in the atmosphere. When a living organism dies, it no longer absorbs new ^{14}C, but the existing one continues to decay. Thus, the half-life can be used to determine the time of death of an organism or the age of organic material from the isotope ratio. The method, also known as ^{14}C dating (pronounced carbon-14 dating), can be used in the time range from 300 to about 60,000 years. Willard Frank Libby developed this dating method in 1946 and received the Nobel Prize in Chemistry for it in 1960.

One problem with the method is that the isotope ratio in the atmosphere is not constant, but changes with solar activity and the carbon cycle. But human activities also have an influence: since industrialisation, large amounts of carbon have entered the atmosphere in the form of CO_2 from fossil fuels. Because these are so old that they no longer contain ^{14}C, its concentration is diluted. From 1945 to 1963, the level of radioactive carbon was also greatly increased by nuclear bomb tests; for this reason, the $^{14}C/^{12}C$ ratio is still higher today than it was before 1945.

Due to these fluctuations, the measurements must be calibrated using other methods. Dendrochronology, for example, which determines the age of trees by means of annual rings, can be used for this purpose.

Excursus 8.2 Snow Line Depression

The determination of the so-called **snow line depression** is made possible by the relatively constant ratio between accumulation and ablation area in balanced glaciers. At the reversal point between advance and retreat, moraines mark a steady state where the mentioned ratio is 2:1. This corresponds to an AAR of 0.67 and means that mass gains are recorded on two thirds of the glacier area.

If the former glacier area is reconstructed on the basis of ice-margin moraines, the area-height distribution of the glacier can be used to determine the altitude of the equilibrium line, which subdivides the glacier into the two areas relevant to the mass balance (Gross et al. 1978). The present snow line is not used as the reference level for the depression, because this would be relatively laborious to determine. For this, one would have to determine the mean equilibrium line altitude, i.e. the highest annual snow line on the glaciers, in each mountain group. In areas where glaciers no longer exist today, it can no longer be determined directly, but only calculated. In order to circumvent these difficulties, the year 1850 is usually used as the reference level, because clear moraines exist from this time in the entire Alpine region.

8.2 Glacial Periods

In the course of the earth's history, there have always been immense climate fluctuations. Whenever one or both poles of the earth are glaciated, scientists speak of an "**ice age**". Strong climate fluctuations also occur within these periods. Phases in which temperatures are significantly below average and larger areas of the mainland are glaciated are called **glacial periods** (alternatively glaciations or glacials). An ice age may contain several glacial periods separated by warmer **interglacial periods** (alternatively interglaciations or interglacials).

Many factors are discussed as causes of cold periods, and it is not yet clear in which cold periods they played a role and to what extent. In any case, plate tectonics is of great importance; it can cause a cooling of the climate in several ways. If large land masses drift towards the pole, they can glaciate and thus additionally reduce the global radiation gain. Snow and ice reflect significantly more solar radiation (60–90%) than land surfaces (5–30%) or water (5%). This positive feedback is called the albedo effect, and it can significantly amplify an initial cooling. In addition, straits close and open as continents move, changing the ocean current pattern and thereby the distribution of thermal energy across the planet. For example, the separation of Antarctica from South America 30–40 million years ago allowed the formation of the Antarctic Circumpolar Currrent, which made the land mass more climatically isolated, eventually leading to the glaciation of the same that continues to this day. Plate tectonics further lead to the formation of mountains, whose enhanced chemical weathering binds CO_2. In phases with reduced CO_2 levels, low solar activity may also be a possible cause.

Once an ice age is underway, the rhythm between cold and warm periods is obviously determined by fluctuations in the Earth's orbital parameters, the so-called Milanković cycles (▶ Excursus 8.3).

If one speaks today of the ice age, then one usually means the last one, which ended only a little more than 10,000 years ago. However, there are also glaciation phases in the history of the earth that are much older and lasted much longer.

Excursus 8.3 Milanković Cycles

The Earth does not orbit around the Sun in a completely stable manner, but wobbles a little. Several parameters overlap with different temporal rhythms: The shape of the Earth's orbit (eccentricity) oscillates back and forth between a more circular and a more elliptical variant every 100,000 years (◻ Fig. 8.3). The inclination of the Earth's axis (Earth's obliquity) is not constant at 23.43° as at present, but varies between 22.1° and 24.5° at intervals of 41,000 years. This mainly affects the radiation budget in the polar regions. The direction of inclination of the rotation axis also changes, similar to a spinning top, with cycles of

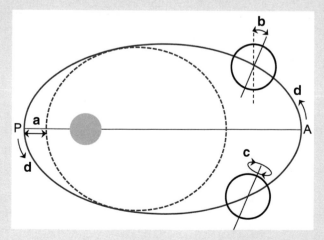

◻ **Fig. 8.3**   Variations of **a** eccentricity, **b** Earth's obliquity, and **c** precession of the Earth's axis. **d** The rotation of perihelion (P) and aphelion (A)

8

about 26,000 years (precession of the Earth's axis). This is superimposed by the precession of the apsides, where the semi-axes rotate once around the Sun in the orbital plane during 112,000 years. The combination of the precessional motions causes the seasonal changes relative to the position on the Earth's orbit to move throughout the calendar year for 21,000 years. Currently, the smallest distance to the sun (perihelion) is in January, and the largest distance (aphelion) is in July. However, this shifts back by one day about every 58 years, so in 10,500 years it will be the other way around. Due to the unequal distribution of land masses in both hemispheres, it makes a difference to global energy gain whether the greatest solar proximity occurs in the northern summer or the southern summer.

The influence of orbital cycles on the world climate and glaciation cycles was first described by James Croll (1821–1890), who was far ahead of his time. Based on this preliminary work, Milutin Milanković worked decades later on his theory, which he titled the "Canon of Insolation and the Ice-Age Problem". The final work was completed in 1941 in Belgrade, where it was almost destroyed during the printing process. But Milanković's theses were also initially rejected by the majority of scientists. It was not until the mid-1970s, when the fluctuations of the great ice sheets could be reconstructed by other methods (▶ Excursus 8.4) and fitted well with Milanković's cycles, that they became generally accepted. Milanković did not live to see the late recognition of his work.

8.2.1 The Older Ice Ages

The oldest traces of large-scale glaciations are two rock horizons in South Africa attributed to the Pongola glaciation 2.9 billion years ago.

For the Huronian (2.4–2.2 billion years B.P.) and Cryogenian (0.7 billion years B.P.) glaciations, the concept of "Snowball Earth" (Kirschvink 1992) was developed. According to this concept, the entire Earth was covered by an ice sheet, and temperatures of −20 °C prevailed even at the equator, leading to global sea ice formation. This concept is highly controversial because organisms could only have survived under a thin ice cover in oceans near the equator.

In the Palaeozoic era, ice ages occurred in the Ordovician/Silurian (450–420 million years B.P.) and Carboniferous/Permian (360–260 million years B.P.) periods. The latter, also called Permo-Carboniferous glaciation, has been demonstrated by field evidence in South Africa on the basis of consolidated moraine deposits ("tillites") (◨ Fig. 8.4).

Africa was then part of the supercontinent Pangaea and was located near the South Pole. The traces of glaciation on the southern continents were among the clues that Alfred Wegener used in 1915 to develop the theory of continental drift, a pioneering precursor of the theory of plate tectonics that is valid today.

▣ Fig. 8.4 Consolidated moraine ("tillite") of the Permo-Carboniferous glaciation in South Africa. The sedimentary rock consists of angular and unstratified fragments of different sizes, indicating deposition by glaciers. (Photo: W. Hagg)

8.2.2 The Pleistocene

The **Pleistocene**, or the "Ice Age" in the narrower sense, refers to first stage of the Quaternary from 2.6 million years B.P. to 11,700 years B.P., which is characterized by an alternation of glacials and interglacials. During the cold periods, about 30% of the continental surface was covered with ice (▣ Fig. 8.5); at present it is just under 10%. Due to the huge masses of water that were bound as ice on the continents, the sea level dropped by more than 100 m during the glacial maxima. Shallow areas of the sea fell dry, the British Isles were as much part of the mainland as, for example, Borneo.

In addition to the present-day ice sheetss of Antarctica and Greenland, there were two other ice sheet: the Scandinavian Ice Sheet included the British Isles and reached as far as Siberia, the Laurentian Ice Sheet covered large parts of Canada, the north of the USA and was possibly even connected to the Greenland ice sheet. Other high mountains were also heavily glaciated, even low mountain ranges like the Bavarian Forest or the Black Forest carried small local glaciers.

Studies of oxygen isotopes from ice cores and marine sediments (▶ Excursus 8.4) now make it possible to reconstruct the temperature record relatively far into the past.

From the ice core of the EPICA project (EPICA = European Project for Ice Coring in Antarctica), which extends to a depth of over 3 km, it was possible to reconstruct the temperatures of the last 800,000 years (▣ Fig. 8.6). While this is extremely impressive, one should be aware that the period covers only one third of the entire Pleistocene.

This temperature profile shows eight glacial cycles in which temperatures in Antarctica were 8–10 °C lower than today. Global cooling was less pronounced and is estimated at 5.8 °C for the last cold period (Schneider von Deimling et al. 2006). It can be seen that during the cold phases the ice sheets grew for over 90,000 years and shrank again in only 10,000 years.

8

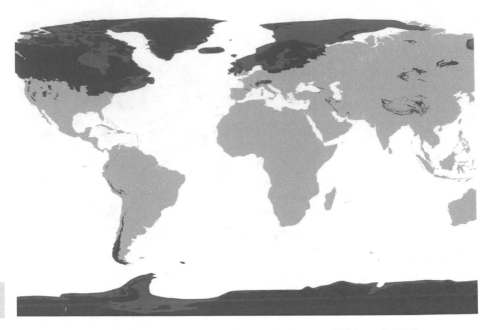

◘ **Fig. 8.5**   The earth in the Würm glacial maximum. (Data source: Ehlers et al. 2011)

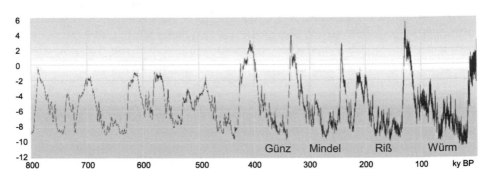

◘ **Fig. 8.6**   Temperature deviations reconstructed from an Antarctic ice core (Dome C) compared to today. (Data source: Jouzel et al. 2007, retrieved from ▶ www.ncdc.noaa.gov on 09/09/2019)

❯ The last ice age is called the Pleistocene, it lasted from 2.6 million to 11,700 years before present and included several glacial periods.

The course of the Pleistocene was reconstructed in the field from the deposits of glaciers (moraines) and their meltwaters (terraces). In the Alpine foothills, sediments from four glaciations (with increasing age: Würm, Riß, Mindel, Günz) were initially identified (Penck and Brückner 1901–1909), which were later extended by two older ones (Donau, Biber) to six (Eberl 1930). A gravel deposit separate from the Mindel terrace is assigned to another, independent glacial period (Haslach) (Schreiner and Ebel 1981), but so far it was only found at one locality in the northeastern Rhine glacier area. Thus, six to seven glaciations in southern Germany, during which the ice masses advanced into the Alpine foothills, can be proven by field evidence. However, the exact number of alpine glacial periods is not known. The advances of the penultimate (Riß) glaciation were mostly the furthest reaching into the foreland north of the Alps and are therefore still well recognizable, even if they have already been somewhat flattened by erosion processes.

With increasing age, field evidence becomes more and more difficult, because the deposits of old glaciations have been largely destroyed by more recent glacier advances. Traces of the older glacials are mainly meltwater deposits, which say nothing about the extent of the glaciers. The supraregional correlation of Pleistocene glacier fluctuations also remains uncertain for the time being.

While southern Germany had the glaciation type of an ice field (within the Alps) with piedmont glaciers, northern Germany was covered by an ice sheet comparable to the present-day inland ice of Greenland (► Chap. 5). The Scandinavian ice sheet reached northern Germany only during the three most recent glacial periods. During the two older glaciations (here called Elster and Saale) the ice reached as far as the German low mountain ranges, during the youngest glacial (here: Weichsel) it deposited its terminal moraines much further towards the Baltic Sea. In both German glaciation areas the Saale/Riß moraines are called older moraines (*"Altmoränen"*) and the moraines of the Weichselian/Würm cold period are called younger moraines (*"Jungmoränen"*). The latter show the greatest freshness of form and often also the greatest heights compared to their surroundings.

> The last two cold periods are called Riß and Würm in southern Germany and Saale and Weichsel in northern Germany. The Alps were covered by a dendritic glacier system, which merged into piedmont glaciers in the foreland that reached as far as the gates of Munich. The Scandinavian ice sheet covered the whole of northern Germany and reached as far as the low mountain ranges.

In the official table of the German Stratigraphic Commission (DSK 2016), only the youngest glacial periods are provided with dates; with increasing age, absolute dating and correlation between Alpine and Nordic glaciation become increasingly difficult.

The chronological sequence of the cold periods and the different regional designations are shown in ◘ Table 8.1.

> With global temperatures about 6 °C lower, the worldwide glacier areas during the Last Glacial Maximum were more than three times those of today, lowering sea level by more than 100 m.

The peak of the last glaciation (Last Glacial Maximum) was reached about 20,000 years ago. After that, the periodic fluctuations of the earth's orbit switched back to warming and, as in the glacials before, the ice masses were degraded

■ Table 8.1   Designation and age of the glacial periods in southern and northern Germany proven by field evidence

Name in southern Germany	Name in North Germany	Age (years before present)
Würm	Vistula	115,000–11,700
Riß	Saale	300,000–126,000
Mindel	Magpie	>380,000
Günz	Menap	<780,000
Danube	Eburon	<1.8 m
Beaver	–	>1.8 m

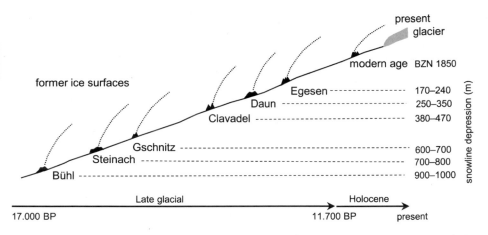

■ Fig. 8.7   Late glacial moraines and associated snowline depressions. Reference level (BZN) is the Little Ice Age peak around 1850. (Modified after Maisch 1982)

much faster than they had formed. In the Alps, the glaciers of the individual valleys were no longer connected across the passes by 16,000 years BC. This means that the dendritic glacier system had been transformed to valley glaciers. The entire period of ice melt from the Last Glacial Maximum to the end of the Würm glaciation 11,700 years ago, when the glaciers reached a modern scale, is called the **Late Glacial.** This deglaciation was not continuous, but was interrupted by several smaller advance phases, called **stadials.** In the Alps, the correlation of the advances between different valleys and mountain groups is not trivial, because the moraines were preserved only in isolated places. ■ Figure 8.7 shows the six known stadials of the Alpine glaciers and the corresponding depressions of the snowline.

In no alpine valley are all glacial stands present as exemplarily as in ■ Fig. 8.7; naturally, the youngest deposits of the Egesen advances (■ Fig. 8.8) are most clearly visible.

Fig. 8.8 Modern moraines from 1850 (white arrows) and Egesen moraines from the Late Glacial (c. 12,500 years B.P.) at Triest Glacier, Valais. (Photo: W. Hagg)

Excursus 8.4 Oxygen Isotope Analysis

This method examines the ratio of the stable oxygen isotopes ^{18}O to ^{16}O, or $\delta^{18}O$ (pronounced delta-O-18) in marine sediments and ice cores. A high value means a high content of heavy ^{18}O isotopes compared to a normal value.

Because water molecules containing the light ^{16}O pass more easily into the gaseous phase, evaporation over the world's oceans enriches ^{18}O in the ocean; seawater is consequently isotopically heavier than the water vapor in the atmosphere (**Fig. 8.9**). While at high air temperatures evaporation is so strong that even some heavy water molecules can still pass into the gaseous phase, at low air temperatures the preference for light oxygen in evaporation is particularly strong. For this reason, in cold phases, the water vapor in the atmosphere and thus the precipitation is isotopically lighter than in warm phases, and the $\delta^{18}O$ value is consequently lower. Because glacial ice is also formed from

this precipitation, ice cores can be used to reconstruct past air temperatures.

Meltwater in the snowpack would smear the isotope signal; therefore, ice cores can only be evaluated at locations where melt never occurs. This limits the method to very high altitude mountainous regions or polar climates. Ice sheets are particularly interesting in this context because they contain very old ice. In addition, isotope ratios in "fossil" air bubbles can be determined in ice cores, although it must be taken into account here that the age of the air does not correspond to that of the ice. Because it can take a long time in high Arctic climates for the firn to become so compressed that the air in the pore space no longer circulates, air bubbles can be significantly younger than the ice by which they are enclosed. In addition to air temperature, ice cores provide other information about the former composition of the atmosphere, e.g. the content

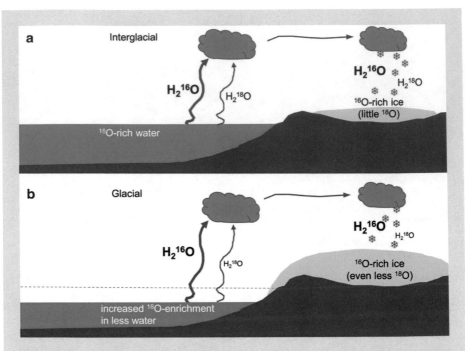

□ **Fig. 8.9** Simplified scheme of the fractionation of oxygen isotopes during (**a**) warm and (**b**) cold periods

of greenhouse gases, or about major events such as volcanic eruptions.

The isotope ratio also changes in the ocean, but here primarily due to the different amount of water: During cold periods, an enormous amount of light water is retained on the continents in the form of glacial ice. During these phases, sea level is lower and the $\delta^{18}O$ in the ocean is much higher than during warm periods. Marine unicellular organisms such as foraminifera then also incorporate more heavy oxygen into their calcareous shells, which are deposited on the seafloor after they die. An isotope analysis of such sediments allows conclusions to be drawn about the extent of conti-

nental glaciation: a high content of ^{18}O indicates a low sea level during a cold period and vice versa. Thus, the signal is exactly opposite to that in the ice cores.

By studying ocean sediments, it became possible to distinguish 103 colder or warmer periods since the beginning of the Pleistocene. These so-called *marine isotope stages* **(MIS)** were numbered backwards, from the current warm period (MIS 1), to the beginning of the Pleistocene (MIS 103). The result is a marine oxygen isotope stratigraphy, and its temporal sequence largely agrees with the Milanković cycles (▶ Excursus 8.3), which was a strong indication of their correctness in the 1970s.

8.3 Glacier Evolution in the Holocene

During the last 11,700 years, the climate was not subject to such strong fluctuations as in the Pleistocene, but it was anything but stable. There were multiple changes between warmer (**optima**) and colder periods (**pessima**), but the magnitude alone, with a maximum of 2 °C, was significantly lower than that between glacial and interglacial. There were probably 12 phases in which the Alpine glaciers were smaller than today (Jörin et al. 2006). This can be proven by tree trunks that melt out of glaciers today and, for example, formed a forest between 8000 and 9000 years B.P. on the Pasterzenboden, where today the longest glacier tongue in Austria is located (Böhm et al. 2007). However, the alpine glaciers never melted completely during the **Holocene,** and there were also phases with significantly larger glacier extents than today. The rough temperature profile of the Holocene is sketched in ◘ Fig. 8.10.

Prominent glacial advances in the **Early Holocene** were the Palü advance around 10,500 years B.P. and the Misox oscillation (also known as 8.2-kiloyear event) around 8200 years B.P., which is attributed to the eruption of a massive glacial lake in North America: Lake Agassiz dammed at the Laurentian Ice Sheet, which melted back northward. The lake contained more water than all of today's lakes combined. When it finally broke through the ice dam in several places and poured up to 70,000 km^3 of water into the Atlantic Ocean over a six-month period, sea level rose by as much as 19 cm (Clarke et al. 2004), and the Gulf Stream, which carries enormous amounts of heat into the North Atlantic, was weakened. This led to cooling in Greenland and Europe (Barber et al. 1999) and caused Alpine glaciers to advance.

The warmest phase was the **Holocene climatic optimum** (7000–6600 years B.P.), during which humans became sedentary and began to practice agriculture. However, even this phase was not continuously warm, as previously assumed, but was interrupted by cooler phases. In a subsequent cooler period lasting a good 1000 years,

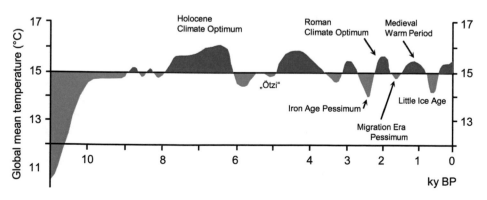

◘ **Fig. 8.10**   Temperature development in the Holocene. (Modified after Schönwiese 1995)

a man died on the Hauslabjoch and thawed out again about 5200 years later as the glacier mummy "Ötzi". Other well-known climatic anomalies of the post-glacial period are the **Roman climatic optimum** (300 BC to 400 AD), during which Hannibal crossed the only moderately glacierized Alps with his elephants. This was followed by the **pessimum of the Migration Period** (450–700 AD), although it is not clear whether it was actually the deterioration of the climate that triggered the migration movements in this period. The **Medieval Climatic Optimum** (750–1500 AD), during which Greenland was settled, is finally followed by the most severe cold period of the Holocene, the so-called **Little Ice Age** (1450–1850 AD). During this period, which is well documented by historical documents, the global mean temperature was 0.8 °C colder than today (Mann et al. 1999); in the Alps a decrease of 2 °C is assumed. The cause is thought to be low solar activity (the Maunder minimum around 1700 AD) and strong volcanic eruptions. These led to the spread of ash into high atmospheric layers, where it reflected part of the solar radiation. The eruption of the Indonesian Tambora in 1815 ejected more than 100 km^3 of vulcanic ash and other pyroclastics and led to crop failures and famine in Europe a year later ("year without summer"). In addition to many phenomena such as frozen canals in the Netherlands, the low temperatures also caused the Alpine glaciers to advance and their total area at their peak around 1850 was well over twice as large as it is today. At that time they had an extremely threatening effect on the mountain dwellers because they destroyed paths, alpine pastures and buildings. Prayer processions were held to stop the growth of the glaciers.

❱ During the Holocene, the temperature fluctuated in a range of about 2 °C, and the glaciers were several times both smaller and larger than today. The last striking peak phase, whose moraines can be found in all high mountains of the earth, was the Little Ice Age with its maximum around 1850 AD, when the Alpine glaciers were more than twice as large as today.

8.4 Current and Future Glacier Retreat

Since the Little Ice Age, global mean temperature has increased by 0.8 °C, to which glaciers worldwide have responded by decreasing their mass, length and area (Paul and Bolch 2019). In the Alps, large glaciers have lost up to 2.5 km of their length, and the total glacier area has decreased by 50–60% from 1850 to 2010. This results in an area loss of 0.34% per year since the Little Ice Age. The warming was not continuous, but was interrupted by cooling periods. During these cooler periods, glaciers also reacted worldwide with advances, in the Alps around 1890, in the 1920s and in the 1970s (Zemp et al. 2006). From 1960 to 1980, a large proportion of Alpine glaciers advanced (Wood 1988); at that time, the media feared a new glacial period. Today, **global dimming** is assumed to be the cause of the cooling. The heavy air pollution at that time, like large volcanic eruptions, reflected a greater proportion of solar radiation than in a clear atmosphere (Wild 2009). From 1973 to 2000, the rate of area loss in the Alps was then 0.6% per year, doubling again in the first decade of this millennium to 1.2% per year (Paul and Bolch 2019).

❯ Even after the Little Ice Age, there were shorter glacier advances in the Alps around 1890 and around 1920, the last occurring in the 1970s when global radiation was reduced by human emissions.

In contrast to the Alpine trend, spectacular glacier advances occurred in Scandinavia and New Zealand in the 1990s (Chinn et al. 2005). The Franz Josef Glacier in the New Zealand Alps advanced by 1.2 km by 1999. But these advances can also be explained by climate warming: Warmer atmosphere can hold more water vapor, leading to higher precipitation and, in cooler regions, more snowfall. In the case of strongly maritime glaciers, this effect overcompensated for a while for the increased melting in summer. Around the turn of the millennium, the mass balance tipped back into negative values as melt once again gained the upper hand. The Franz Josef Glacier has also retreated strongly again in the meantime (▶ Chap. 6).

The current glacier retreat is a worldwide signal, the only exception currently being found in the Karakoram and the neighbouring Kunlun Shan and Pamir mountains. Here, glacier masses are stable or even growing slightly (Hewitt 2005; Gardelle et al. 2013; Bolch et al. 2017). The reason could be the same as in the 1990s in Norway and New Zealand: increased snowfall due to higher temperatures that (still) outweighs the greater melt. Recent research shows that intensification of irrigation in the fields of the surrounding lowlands causes an increase in summer snowfall and a decrease in solar radiation, which counteract climate warming (de Kok et al. 2018). However, this so-called **Karakoram anomaly** will also be only a temporary phenomenon in the course of prolonged climate warming, similar to the advances in New Zealand and Norway (Farinotti et al. 2020).

❯ The only region of the world with regional glacier growth is currently the Karakoram, where the mass increases are probably due to temporarily higher snowfall.

The World Glacier Monitoring Service (WGMS) in Zurich collects data on glaciers and their changes worldwide. Glaciers with a continuous measurement series of over 30 years are referred to as reference glaciers. The regionally averaged decadal mass balance of these glaciers was −214 mm w.e. a^{-1} in the 1980s and already −499 mm w.e. a^{-1} in the 1990s. From 2010 to 2019, the mean value was −924 mm w.e. a^{-1}, the peak value of −1177 mm w.e. a^{-1} was reached in 2019 (WGMS 2021). The time course of this mean value into the current decade (◻ Fig. 8.11) makes the trend more than clear: while in the old millennium mean values of −600 mm w.e. were strongly negative years, the same value in the new millennium represents a rather weakly negative year.

Measurements of the mass balance using the glaciological method began in Sweden in 1945, since when glaciers have lost almost 30 m w.e. of mass (WGMS 2021). Due to the fact that the adjustment of glacier area and geometry is delayed, glaciers are currently lagging behind climate development. This means that even if warming were to stop immediately, the shrinkage would continue for years and, in

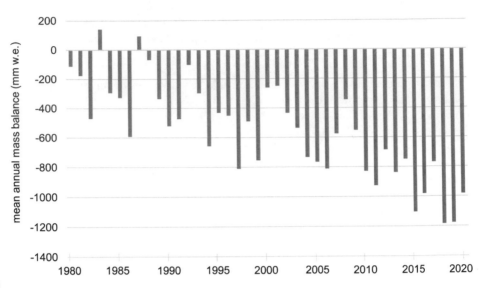

 Fig. 8.11   Mean mass balance of the World Glacier Monitoring Service reference glaciers. (Data source: WGMS 2021)

the case of large glaciers, for decades, and glaciers would have to lose a further 25–65% of their area, depending on the region, in order to adapt to today's climate (Zemp et al. 2015).

> In the vast majority of regions, glacier retreat has accelerated in recent decades. The sluggishly reacting glaciers are lagging behind the climate development, otherwise they would already have shrunk even more.

The future development of glaciers depends on several factors; the most important are climate, the distribution of ice thickness and ice movement. While the current ice thickness can be measured reasonably well or derived from the slope of the glacier surface, climate scenarios are subject to relatively large uncertainties and the simulation of ice dynamics poses an additional challenge. Furthermore, there are other effects such as increasing debris cover that affect glacier development and are very difficult to predict. All this complicates forecasts of glacier development and causes uncertainties in the corresponding scenarios. While the general future trend is clear, the rate of glacier retreat varies widely, for example depending on the underlying climate scenario. In the most recent Assessment Report of the IPCC (Intergovernmental Panel on Climate Change), five different so-called "shared socioeconomic pathways" are used to describe the development of greenhouse gas emissions with different climate policies in the twentyfirst century. The most optimistic scenario assumes a significant reduction in emissions, resulting in a warming of about 1.0–1.8 °C relative to 1850–1900. Under the most pessimistic assumptions, models calculate a temperature increase of 3.3–5.7 °C (IPCC 2021). According to a recent study, a warming in the latter range leads to a reduction of the current glacier area in the Alps by about 65–90% by 2100 (Zekollari et al. 2019).

The mountains will therefore not be completely ice-free by the end of the century, but assuming a strong increase in greenhouse gases, only small remnants of today's largest glaciers will remain.

> By the end of the century, in all probability, only the remains of those glaciers that are still among the largest today will be found in the Alps.

8.5 Consequences of Glacier Retreat

The consequences of glacier retreat can be considered at three scales: local, regional and global.

8.5.1 Local Consequences

Locally, the disappearance of mountain glaciers can lead to an increase in natural hazards. In places where glaciers have steepened rock sections through their erosive activity, slope instabilities arise after melting. Due to the lack of support from the ice, rockfalls can be triggered in this way. This has happened, for example, at the Lower Grindelwald Glacier, where the downwasting of the ice surface since 1860 by about 300 m caused a massive rockfall in 2006 (Oppikofer et al. 2008). The shrinkage of glaciers often creates large areas of debris where the terrain was previously ice-covered. This unconsolidated and unvegetated loose material is now exposed to the action of the atmosphere and can be mobilized during heavy rainfall events. This creates debris flows, a mixture of rock and water that has flow properties similar to concrete and a high potential for destruction due to its high density. Chiarle et al. (2007) compared debris flows from glacier forelands over the past 25 years with historical data and conclude that the frequency of these events is increasing due to glacier retreat.

> Due to glacier recession, the danger of rockfalls and debris flows is increasing locally.

8.5.2 Regional Consequences

Rivers with glaciers in their catchments show regional effects of glacier retreat that affect streamflow on multiple temporal scales.

The rise in the equilibrium line reduces the size of the firn areas that can temporarily store water as hydrological buffers (▶ Chap. 7). This increases the diurnal fluctuations in runoff, which are already high in heavily glaciated catchments.

Glaciers are seasonal water reservoirs that retain winter precipitation in solid form and release it as meltwater during the ablation period until late summer. This seasonal redistribution is particularly important in areas with dry summers, as it ensures water availability during dry-hot periods even after the snowpack has been consumed. In large parts of Central Asia, irrigated agriculture would not be possible without the existence of glaciers (Hagg 2018).

However, glaciers distribute water not only within a year, but also over several years due to changes in their size: in cool phases, they build up their reserves and grow; in hot phases, they release the surplus back into the rivers and lose mass and volume. They thus guarantee a minimum streamflow, which can be of enormous importance for hydropower, navigation and industry. Rivers with glaciers in their catchment area always carry water that, depending on the weather, comes either from rain or from melt.

In the case of climate warming, the mass balance of the glaciers tips into the negative range over several years. The consequences for ice melt are shown in ◘ Fig. 8.12.

The additional melt due to the rise of the equilibrium line and the depletion of ice reserves initially creates an oversupply of water (phase b in ◘ Fig. 8.12). However, with a time lag, glacier areas then begin to shrink, slowing the increase in ice melt. At some point, a maximum of melt and runoff is reached, from where the area shrinkage outweighs the increased melt rates and water availability decreases. Initially it is still above the initial level (phase c), but eventually it falls below it (phase d). When glaciers disappear completely after a strong warming, ice melt approaches zero. If the glacier areas adapt to the new climate, then the ice melt also finds a new equilibrium (phase e). At present, in the glaciated high mountains of

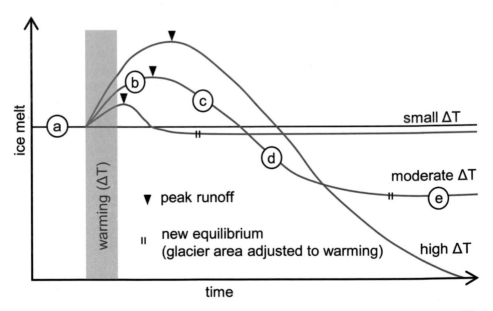

◘ **Fig. 8.12**   Change in ice melt following climate warming. Typical phases are: Phase **a** = equilibrium state in a stable climate, Phase **b** = increased runoff during and after warming, Phase **c** = decreasing but still above initial level, Phase **d** = decreasing runoff, below initial level, Phase **e** = new equilibrium. (Modified after Hagg 2018)

the Earth, we are still in the phase of an increased water supply compared to the equilibrium state. The tipping point has probably already been passed in less heavily glaciated mountains such as the Alps today, while runoff in heavily glaciated mountains will continue to increase in the near future (Duethmann et al. 2016; Hagg et al. 2018). However, the continued glacier retreat that can currently be observed will ultimately lead to a worldwide decrease and disappearance of both hydrological redistribution effects of glaciers, seasonal and multi-annual. This causes stronger fluctuations in water flow, which is then only controlled by precipitation. Stronger spring floods due to more intensive snowmelt will be followed by more pronounced summer low-water periods in the future (Hagg et al. 2007; Sorg et al. 2014).

❯ In glacier-fed rivers, after an initial increase in discharge, glacier retreat causes a shift of water reserves to spring, greater fluctuations from year to year, and an accumulation of summer water deficits.

At the Vernagtferner in the Ötztal Alps, the hydrological consequences of glacier retreat have long been clearly visible. Discharge fluctuations within a day in the new millennium are many times higher than typical values from the 1970s (Braun et al. 2013). Annual streamflow has also doubled from 1974 to 2011, with the largest increases occurring in June and May, while streamflow actually decreases from July to September (Escher-Vetter and Reinwarth 2013). This illustrates the seasonal redistribution that is currently taking place and is far from complete.

8.5.3 Global Consequences

On a global scale, sea level rise is an important and dramatic consequence of glacier wastage for humans. Mountain glaciers and ice caps, i.e. all glacial ice excluding the two ice sheets, currently cover an area of about 700,000 km^2 (RGI Consortium 2017). From 1961 to 2016, their melting contributed 25–30% to current sea level rise (2.6–2.9 mm a^{-1}), about as much as Greenland and significantly more than Antarctica (Zemp et al. 2019). Antarctica's contribution is smaller because this ice sheet has so far been relatively little affected by mass losses. However, this picture will be reversed in the future, as the mountain glaciers are already melting at a very high rate and there is comparatively little mass left, while the giant ice sheets have just started to fill up the oceans. A complete melting of the mountain glaciers would raise sea level by 0.4 m (Huss and Farinotti 2012), while the sea level equivalent of Greenland is 7.3 m and that of Antarctica 56.6 m (Lemke et al. 2007).

❯ If all mountain glaciers melt, sea level will rise by 0.4 m; for the Greenland ice sheet, the value is 7.3 m. The Antarctic ice sheet, with a sea-level equivalent of 56.7 m, would significantly alter the Earth's coastlines. However, complete melting of the inland ice would take millennia even with a very strong rise in temperature.

References

Armstrong RA (2016) Lichenometric dating (lichenometry) and the biology of the lichen genus rhizocarpon: challenges and future directions. Geogr Ann Se A Phys Geogr 98(3):183–206. https://doi.org/10.1111/geoa.12130

Barber D, Dyke A, Hillaire-Marcel C et al (1999) Forcing of the cold event of 8, 200 years ago by catastrophic drainage of Laurentide lakes. Nature 400:344–348. https://doi.org/10.1038/22504

Beschel R (1950) Flechten als Altersmaßstab rezenter Moränen. Z Gletscherk Glazialgeol 1:152–161

Böhm R, Schöner W, Auer I, Hynek B, Kroisleitner C, Weyss G (2007) Gletscher im Klimawandel. Zentralanstalt für Meteorologie und Geodynamik, Wien

Bolch T, Pieczonka T, Mukherjee K, Shea J (2017) Brief communication: glaciers in the Hunza catchment (Karakoram) have been nearly in balance since the 1970s. Cryosphere 11(1):531–539. https://doi.org/10.5194/tc-11-531-2017

Braun L, Reinwarth O, Weber M (2013) Der Vernagtferner als Objekt der Gletscherforschung. Z Gletscherk Glazialgeol 45(46):85–104

Chiarle M, Iannotti S, Mortara G, Deline P (2007) Recent debris flow occurrences associated with glaciers in the Alps. Glob Planet Chang 56:123–136

Chinn TJH, Winkler S, Salinger MJ, Haakensen N (2005) Recent glacier advances in Norway and New Zealand: a comparison of their glaciological and meteorological causes. Geogr Ann Ser A Phys Geogr 87(1):141–157. https://doi.org/10.1111/j.0435-3676.2005.00249.x

Clarke GKC, Leverington DW, Teller JT, Dyke AS (2004) Paleohydraulics of the last outburst flood from glacial Lake Agassiz and the 8200 BP cold event. Quat Sci Rev 23:389–407. https://doi.org/10.1016/j.quascirev.2003.06.004

de Kok RJ, Tuinenburg OA, Bonekamp PNJ, Immerzeel WW (2018) Irrigation as a potential driver for anomalous glacier behavior in High Mountain Asia. Geophys Res Lett 45:2047–2054. https://doi.org/10.1002/2017GL076158

DSK (2016) Stratigraphische Tabelle von Deutschland. Deutsche Stratigraphische Kommission (Hrsg; Koordination und Gestaltung: Menning M, Hendrich A). Deutsches GeoForschungsZentrum, Potsdam. ISBN 978-3-9816597-7-1

Duethmann D, Menz C, Jiang T, Vorogushyn S (2016) Projections for headwater catchments of the Tarim River reveal glacier retreat and decreasing surface water availability but uncertainties are large. Environ Res Lett 11(5):054024

Eberl B (1930) Die Eiszeitfolge im nördlichen Alpenvorlande. Ihr Ablauf, ihre Chronologie auf Grund der Aufnahmen des Lech-und Illergletschers. Benno Filser, Augsburg

Ehlers J, Gibbard PL, Hughes PD (2011) Quaternary glaciations – extent and chronology. http://booksite.elsevier.com/9780444534477/. Accessed on 02.02.2020

Escher-Vetter H, Reinwarth O (2013) Meteorologische und hydrologische Registrierungen an der Pegelstation Vernagtbach – Charakteristika und Trends ausgewählter Parameter. Z Gletscherk Glazialgeol 45(46):117–128

Farinotti D, Immerzeel WW, de Kok RJ et al (2020) Manifestations and mechanisms of the Karakoram glacier anomaly. Nat Geosci 13:8–16. https://doi.org/10.1038/s41561-019-0513-5

Furrer G, Holzhauser H (1984) Gletscher- und klimageschichtliche Auswertung fossiler Hölzer. Z Geomorphol Neue Folge Suppl 50:117–136

Gardelle J, Berthier E, Arnaud Y, Kääb A (2013) Region-wide glacier mass balances over the Pamir-Karakoram-Himalaya during 1999–2011. Cryosphere 7(4):1263–1286. https://doi.org/10.5194/tc-7-1263-2013

Gross G, Kerschner H, Patzelt G (1978) Methodische Untersuchungen über die Schneegrenze in alpinen Gletschergebieten. Z Gletscherk Glazialgeol 12(2):223–251

Hagg W (2018) Water from the mountains of greater Central Asia: a resource under threat. In: Squires V, Qi L (eds) Sustainable land management in greater Central Asia. Routledge, London/New York, pp 237–248

Hagg W, Braun LN, Kuhn M, Nesgaard TI (2007) Modelling of hydrological response to climate change in glacierized central Asian catchments. J Hydrol 332:40–53

Hagg W, Mayr E, Mannig B, Reyers M, Schubert D, Pinto J, Peter J, Pieczonka T, Bolch T, Paeth H, Mayer C (2018) Future climate change and its impact on runoff generation from the debris-covered Inylchek Glaciers, Central Tian Shan. Kyrgyzstan Water 10:1513. https://doi.org/10.3390/w10111513

Hewitt K (2005) The Karakoram anomaly? Glacier expansion and the „elevation effect". Karakoram Himalaya Mt Res Dev 25:332–340

Holzhauser H (2009) Auf dem Holzweg zur Gletschergeschichte. In: Hallers Landschaften und Gletscher. Beiträge zu den Veranstaltungen der Akademien der Wissenschaften Schweiz 2008 zum Jubiläumsjahr „Haller 300". Sonderdruck aus den Mitteilungen der Naturforschenden Gesellschaft in Bern. Neue Folge 66:173–208

Holzhauser H, Magny M, Zumbühl HJ (2005) Glacier and lake-level variations in west-central Europe over the last 3500 years. The Holocene 15(6):789–801

Huss M, Farinotti D (2012) Distributed ice thickness and volume of all glaciers around the globe. J Geophys Res 117:F04010

IPCC (2021) Climate change 2021: the physical science basis. In: Masson-Delmotte V, Zhai P, Pirani A, Connors SL, Péan C, Berger S, Caud N, Chen Y, Goldfarb L, Gomis MI, Huang M, Leitzell K, Lonnoy E, Matthews JBR, Maycock TK, Waterfield T, Yelekçi O, Yu R, Zhou B (eds) Contribution of working group I to the sixth assessment report of the intergovernmental panel on climate change. Cambridge University Press. In Press

Jörin U, Stocker TF, Schlüchter C (2006) Multicentury glacier fluctuations in the Swiss Alps during the Holocene. The Holocene 16(5):697–704

Jouzel J, Masson-Delmotte V, Cattani O, Dreyfus G, Falourd S, Hoffmann G, Minster B, Nouet J, Barnola JM, Chappellaz J, Fischer H, Gallet JC, Johnsen S, Leuenberger M, Loulergue L, Luethi D, Oerter H, Parrenin F, Raisbeck G, Raynaud D, Schilt A, Schwander J, Selmo E, Souchez R, Spahni R, Stauffer B, Steffensen JP, Stenni B, Stocker TF, Tison JL, Werner M, Wolff EW (2007) Orbital and millennial Antarctic climate variability over the past 800,000 years. Science 317(5839):793–797, 10 August. Ncdc.noaa.gov. Accessed on 09.09.2018

Kirschvink JL (1992) Late proterozoic low-latitude global glaciation: the snowball earth. In: Schopf JW, Klein C (eds) The proterozoic biosphere. Cambridge University Press, Cambridge, pp 51–52

Lemke P, Ren J, Alley RB, Allison I, Carrasco J, Flato G, Fujii Y, Kaser G, Mote P, Thomas RH, Zhang T (2007) Observations: changes in snow, ice and frozen ground. In: Solomon S, Qin D, Manning M, Chen Z, Marquis M, Averyt KB, Tignor M, Miller HL (eds) Climate change 2007: the physical science basis. Contribution of working group I to the fourth assessment report of the intergovernmental panel on climate change. Cambridge University Press, Cambridge, UK/New York

Locke WW, Andrews JT, Webber PJ (1979) A manual for lichenometry. Br Geomorphol Res Group Tech Bull 26:1–47

Maisch M (1982) Zur Gletscher- und Klimageschichte des alpinen Spätglazials. Geog Helv 82(2):93–104

Mann ME, Bradley RS, Hughes MK (1999) Northern hemisphere temperatures during the past millennium: inferences, uncertainties, and limitations. Geophys Res Lett 26:759–762

Oppikofer T, Jaboyedoff M, Keusen HR (2008) Collapse at the eastern Eiger flank in the Swiss Alps. Nat Geosci 1:531–535. https://doi.org/10.1038/ngeo258

Paul F, Bolch T (2019) Glacier changes since the little ice age. In: Heckmann T, Morche D (eds) Geomorphology of proglacial systems. Geography of the physical environment. Springer, Cham

Penck A, Brückner E (1901–1909) Die Alpen Im Eiszeitalter, Bd 3. Tauchnitze, Leipzig

RGI Consortium Randolph Glacier Inventory (v.6.0): A dataset of global glacier outlines. Global land ice measurements from space, Boulder, Colorado USA (RGI technical report, 2017). https://doi.org/10.7265/N5-RGI-60

Schneider von Deimling T, Ganopolski A, Held H, Rahmstorf S (2006) How cold was the last glacial maximum? Geophys Res Lett 33:L14709. https://doi.org/10.1029/2006GL026484

Schönwiese C-D (1995) Klimaänderungen. Springer, Berlin/Heidelberg

Schreiner A, Ebel R (1981) Quartärgeologische Untersuchungen in der Umgebung von Interglazialvorkommen im östlichen Rheingletschergebiet (Baden-Württemberg). Geologisches Jahrbuch/A59. Schweizerbart, Hannover

Sorg A, Russ M, Rohrer M, Stoffel M (2014) The days of plenty might soon be over in glacierized Central Asian catchments. Environ Res Lett 9:104018. (8 pp). https://doi.org/10.1088/1748-9326/9/10/104018

WGMS (2021) Latest glacier mass balance data. https://wgms.ch/latest-glacier-mass-balance-data/. Accessed on 23.12.2021

Wild M (2009) Global dimming and brightening: a review. J Geophys Res 114:D00d16. https://doi.org/10.1029/2008jd011470

Winkler S (2000) Der „Schmidt-Hammer" als geochronologische Methode – Anwendungsmöglichkeiten und Problematik aufgezeigt an Beispielen aus Neuseeland und Norwegen. Trierer Geogr Stud 23:123–146

Winkler S (2005) The ‚Schmidt hammer' as a relative-age dating technique: potential and limitations of its application on Holocene moraines in Mt Cook National Park, Southern Alps, New Zealand. N Z J Geol Geophys 48:105–116

Wood FB (1988) Global alpine glacier trends, 1960s to 1980s. Arct Alp Res 20(4):404–413

Zekollari H, Huss M, Farinotti D (2019) Modelling the future evolution of glaciers in the European Alps under the EURO-CORDEX RCM ensemble. Cryosphere 13:1125–1146. https://doi.org/10.5194/tc-13-1125-2019

Zemp M, Paul F, Hoelzle M, Haeberli W (2006) Glacier fluctuations in the European Alps 1850–2000: an overview and spatiotemporal analysis of available data. In: Orlove B, Wiegandt E, Luckman B (eds) The darkening peaks: glacial retreat in scientific and social context. University of California Press, Berkley/Los Angeles, pp 152–167

Zemp M, Frey H, Gärtner-Roer I, Nussbaumer SU, Hoelzle M, Paul F, Haeberli W, Denzinger F et al (2015) Historically unprecedented global glacier decline in the early 21st century. J Glaciol 61(228):745–762

Zemp M, Huss M, Thibert E, Eckert N, McNabb R, Huber J, Barandun M, Machguth H, Nussbaumer SU, Gärtner Roer I, Thomson L, Paul F, Maussion F, Kutuzov S, Cogley JG (2019) Global glacier mass changes and their contributions to sea-level rise from 1961 to 2016. Nature 568:382–386. https://doi.org/10.1038/s41586-019-1071-0

Zumbühl HJ, Nussbaumer SU (2018) Little ice age glacier history of the Central and Western Alps from pictorial documents. https://doi.org/10.18172/cig.3363

8

Glacial Hazards

Contents

Overview
Natural hazards are omnipresent in high mountains. Two threats come directly from glaciers: Ice avalanches and glacial lake outbursts. The two phenomena can be divided into different types in terms of their origin, for which there are equally impressive and tragic examples from the Alps and from other high mountains. Depending on the triggering mechanism and the course of the event, the population in the valleys has various options for risk minimisation and disaster preparedness.

The title of this chapter may prompt many readers to think of crevasses and other dangers that glaciers can pose to mountaineers. However, these are not the subject of consideration here, but rather those hazards that emanate from the glacier to its surroundings and to inhabited valleys. In natural risk research, the term "natural hazard" refers to a hypothetical event with potential damage to humans, whereby both property damage and personal injury are meant here. A "natural phenomenon", on the other hand, has already occurred but has only had an impact on the natural environment, which can occur particularly in areas that are uninhabited and unused by humans. If, on the other hand, an event has already resulted in enormous damage to humans, it is referred to as a "natural disaster" (Felgentreff and Glade 2008). According to this view, natural hazards only exist for humans and natural disasters are only catastrophic for humans.

High mountains are particularly susceptible to natural hazards for various reasons. The large differences in altitude and the steep relief enable landslides and favour floods and debris flows. The presence of a winter snow cover brings with it the danger of (snow) avalanches which, despite technical protective measures, repeatedly lead to major damage. In the following, however, only natural hazards that originate directly from glaciers will be described, i.e. ice avalanches and glacial lake outburst floods.

9.1 Ice Avalanches

9.1.1 Definition and Classification

The term **ice avalanche** (synonyms: ice fall, glacial avalanche) refers to a mass of ice that has broken away from a glacier and is falling down to the valley. Very large and rare ice avalanches with a volume of more than 1 million m^3 are also called glacial avalanches. Analogous to landslides, a spatial distinction is made between the starting zone (◘ Fig. 9.1), where the event originates, the avalanche path and the deposition area.

Ice avalanches can be divided into two basic types (◘ Fig. 9.2): In the "ramp" type, the breaking-off part of the glacier lies on the bedrock before the event, while the "cliff" type denotes ice avalanches that break off at steep cliff and may have only relatively small ice volumes (Haefeli 1966). In the case of the "ramp" type, the thermal regime of the ice determines the further subdivision: in warm

◘ Fig. 9.1 Hanging glacier in the Tienshan, where an ice avalanche apparently broke loose not long ago. (Photo: David Kriegel)

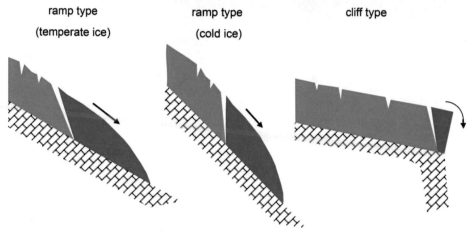

◘ Fig. 9.2 Three types of ice avalanches. (Modified after Alean 1985)

ice, which is close to the pressure melting point and where a film of meltwater between ice and rock allows basal sliding, ice avalanches are triggered from a gradient of 25°. This subtype occurs mainly during the ablation period, i.e. in summer. Cold ice, on the other hand, can be frozen to the bedrock on steep rocky terrain of 45° slope and more. Ice avalanches occur here irregularly and independent of the season.

> ❯ Cold ice can still adhere to the rock at very steep inclinations of 45° and more. In temperate ice, the basal meltwater film considerably reduces friction, which is why ice avalanches occur here from as little as 25°.

9.1.2 Examples

The largest documented glacial avalanche in the Alps occurred in 1895 on the Altels, a mountain in the Swiss canton of Valais, whose smooth northwestern flank was heavily glaciated in the nineteenth century (◘ Fig. 9.3).

☐ Fig. 9.3 The Altels one year before and shortly after the disaster. (Photographer: Paul Montandon. Alpine Museum of Switzerland, Bern)

The ice thickness measured at the break-off edge after the event was 40 m. At 5 o'clock in the morning on 11th September (sic!), 4.5 million m³ of ice (roughly equivalent to the volume of 3000–4000 single-family houses or 1.7 Cheops pyramids) broke loose and plunged down a drop of 1440 m at a speed of 450 km h⁻¹, only to rush 320 m up the opposite slope (Heim 1895; ☐ Fig. 9.4).

The 5 m thick deposits covered an area of about 1 km². Outside these deposits, one hut was destroyed by the shock wave alone, killing six people. Other fatalities included a dog, a mule, nine pigs and 158 cattle. The cattle were also hurled by the wind pressure with a throw of up to 350 m over a distance of up to 1000 m, "they flew before the avalanche like autumn leaves before the storm" (Heim 1895).

Large snow and ice avalanches are preceded by an enormous pressure wave that can knock down trees and be fatal for people. The danger zone here is therefore even greater than the obvious one affected by the avalanche itself.

The glacier collapse lasted about 1 min and could still be heard in Beatenberg, more than 30 km away. The cause of this tremendous event was a warming of the glacier bed from cold to temperate ice and an unfavourable mass distribution. Due to the preceding advance during the Little Ice Age, the glacier front was still thickened during the subsequent retreat (Röthlisberger 1981). Due to the small size of the glacier, a repetition of the event, at least on a similar scale, is no longer possible.

An example of a more recent natural disaster is the collapse on the Allalin glacier, also located in the canton of Valais. There, on 30th August 1965, the construction site for a reservoir including residential barracks was buried 10 m high by ice that had broken off from the glacier tongue 450 m higher up (☐ Fig. 9.5). The total volume of the ice mass was 2 million m³ and it buried 88 people, for whom all help came too late. In 1999 and 2000, smaller ice avalanches occurred again at this location, but the glacier is monitored intensively and danger areas are closed if necessary.

In 2017, a large ice avalanche with a volume of 400,000 m³ occurred on the Swiss Trift Glacier, resulting in no personal injuries but briefly evacuating part of a village.

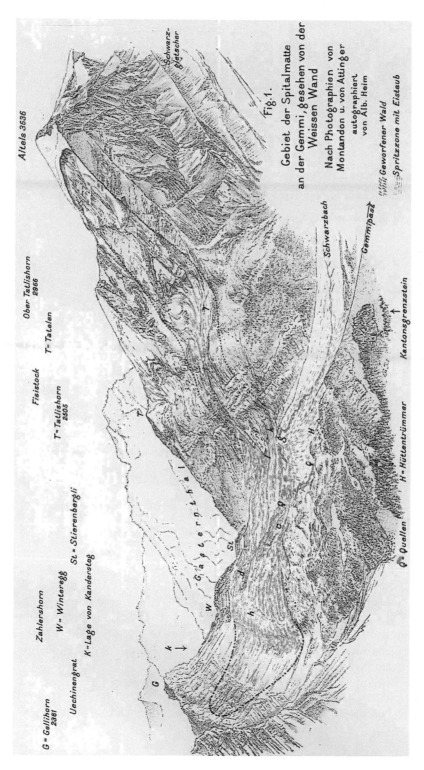

◻ Fig. 9.4 Area of the Spitalmatte after the glacier collapse. (From Heim 1895, with kind permission from the Zürcherische Naturforschende Gesellschaft)

9

■ Fig. 9.5 The construction site of the Mattmark reservoir before and after the disaster. (Photos: Peter Kasser, archive VAW/ETH Zurich, with kind permission)

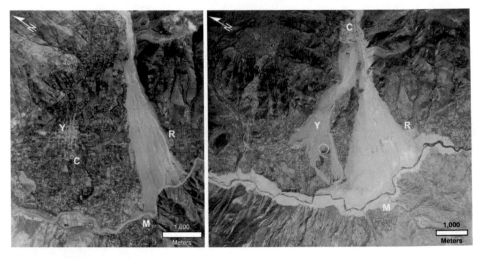

■ Fig. 9.6 Deposits from the 1962 (left) and 1970 (right) mudslides. Letters denote localities, Y is the locality of Yungay. (From Evans et al. 2009b, courtesy © Elsevier AG 2009, all rights reserved)

Also outside the Alps there are spectacular and at the same time tragic examples of ice avalanches and glacier collapses. On Nevado Huascarán, a mountain in Peru, catastrophic ice avalanches have twice broken loose, taking a lot of debris with them. Melting processes and entry into a stream bed gradually liquefied the mixture. The resulting mud-rich debris flows largely buried and destroyed the town of Yungay and other settlements (■ Fig. 9.6). In 1962, about 1000 people died, and in 1970, about 7000 people died (Evans et al. 2009a), although casualty figures are even much higher, depending on the source.

Since the Little Ice Age, the Kolka glacier in the Caucasus has shown a surge behaviour with a periodicity of about 67 years (1835, 1902, 1970). In 2003, an exceptional event occurred outside of this cycle in which the glacier not only

◘ Fig. 9.7 Top: The ice and rock avalanche of Kolka glacier destroyed vegetation up to heights of 140 m above the valley floor (left). Bottom: Above the narrows where the mass stopped, it dammed a lake. (Photos: Gennady Nosenko, September 2003)

advanced but completely detached from its bed and, as a combined ice and rock avalanche, traveled a distance of 19 km through the Genaldon Valley at a speed of 180 km h^{-1} (Evans et al. 2009b), devastating a large area (◘ Fig. 9.7). At one narrow point, the mass was slowed down, but the squeezed-out water destroyed another 15 km of the valley floor as a debris flow. A village was buried and a total of 125 people lost their lives.

Due to the sliding rather than falling form of movement, the event does not represent a glacier avalanche in the strict sense. Since there is no comparable catastrophe, Evans et al. (2009b) propose the term "Kolka-type behaviour", which describes a maximum conceivable glacier instability. Several triggers for the event are discussed (Kotlyakov et al. 2004; Evans et al. 2009b): unusually high melt rates throughout the summer and heavy rainfall shortly before the disaster led to water supersaturation in the glacier and possibly to floating of the ice masses. Sulfur odor after the disaster indicated increased subglacial melting due to thermal water seepage. The direct trigger was then probably a weak earthquake or an ice avalanche caused by it, which fell down from the steeper rock faces onto the tongue.

9.1.3 Risk Management

The possibilities for human action begin with an inventory of dangerous glaciers, as was carried out in Switzerland by the Laboratory of Hydraulics, Hydrology and Glaciology (VAW) at the Swiss Federal Institute of Technology in Zurich (ETH) (Raymond et al. 2003). According to this data collection, in the period from 1595 to 2003 there were 17 ice avalanches with fatalities (253 deaths in total) and 48 ice avalanches (on average every 9 years) resulting in property damage on Swiss territory.

The next step is the monitoring of dangerous glaciers. In 2003, there were 31 glaciers in Switzerland with a risk of ice avalanches, 12 of which were permanently monitored. Changes in geometry and crevasse formation as well as changes in surface velocity were recorded, also by means of GPS receivers.

The Glariskalp-Alcotra project (2020) records 47 historical and current ice avalanches with 17 fatalities in the French-Italian border area of the Mont Blanc massif. In September 2019, the Planpincieux glacier in the Valle d'Aosta came under media scrutiny here because of a large crevasse below which the tongue was moving at up to 70 cm per day, seven to 14 times its normal speed. It threatened to break off 250,000 m³ of ice. The temperate glacier has been monitored since 2015 with an automated camera that takes a photo every hour (Dematteis et al. 2018). It was found that strong, short-term ice accelerations and 87 ice break-offs already occurred during the summers of 2015–2017, including eight major ones with volumes of 5000–60,000 m³. Here, from an ice movement of 30 cm per day, the probability of a break-off increases to 90% (Giordan et al. 2020). A large ice avalanche would take only 80 s to reach the valley floor. In September 2019, two roads were closed for this reason, but the major event did not (yet) occur.

The simulation of ice avalanches by means of mathematical models ranges from the estimation of travel distances to the simulation of the fall movement, although the low data availability here does not yet permit operational use, for example for the designation of danger zones. Predictions are theoretically possible on the basis of the characteristic progressive acceleration (Röthlisberger 1981) of the breaking-off part of the glacier before the event. On this basis, the prediction of an ice avalanche on the Weisshorn in the canton of Valais was already successful in 1973 with an error of only 4 days (Flotron 1977). However, concrete forecasts will probably only be possible in the rarest of cases in the future. At temperate ramp glaciers, an acceleration can also be caused by variations in basal sliding, which makes prediction more difficult (Pralong and Funk 2006).

9.2 Glacial Lake Outburst Floods

9.2.1 Classification and Examples

Glacal lake **outburst floods (GLOFs)** are spontaneous discharges of glacial lakes. These lead to floods that do not differ in their effects from other floods; only the cause is directly linked to the action of the glaciers. In his classification of glacial lakes, Schweizer (1957) distinguishes between water accumulations in a glacier (water pockets), lakes on a glacier (supraglacial lakes) and dammed glacial lakes. In the latter case, it also depends on whether ice or moraine acts as a barrier, and finally the location of the lake in relation to the glacier situation in the main and secondary valley determines which type is present (**◘** Fig. 9.8).

In the case of **moraine-dammed lakes,** the simplest and most common case is that the terminal moraine of a glacier dams up its own runoff to form a proglacial lake. Hollow forms are currently appearing everywhere in the glacier forefield due to global glacier retreat, and these can fill with water, which is why an increase in this type of lake is being recorded in many places. In Bhutan, Nagai et al. (2017) mapped 733 glacial lakes, 24 of which are classified as hazardous (Mool et al. 2001). According to Schweizer's (1957) classification, the proglacial lake type is called "Cohup", named after Laguna Cohup in Peru, which had an outburst in 1941 (**◘** Fig. 9.9), causing 7000 deaths in the city of Huaraz. This is one of the largest natural disasters caused by glaciers.

A counterexample is Nostetuko Lake in British Columbia. Here, a proglacial lake outburst occurred in 1983 with flood peaks of 10,000 $m^3 \ s^{-1}$ (Clague and Evans 2000), which is more than the average discharge of the Rhine plus the Danube at their respective mouths. However, because the region is uninhabited, no human damage occurred, which classifies the eruption as a natural phenomenon in the sense of natural risk research and not as a natural disaster.

If the lateral moraine of a side valley glacier dams up the outflow in the main valley, this is known as the "Mattmark" type, named after the lake of the same name in the Saas Valley, which had catastrophic outbursts in 1589, 1633, 1680 and 1772. In 1680, 18 houses were destroyed in Visp, 29 km away. Today, the lake can no longer burst out because its dam has since been artificially reinforced (**◘** Fig. 9.10).

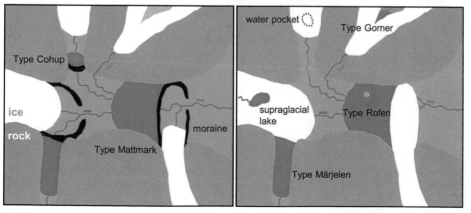

a moraine-dammed lakes   **b** ice-dammed lakes

◘ Fig. 9.8 Schematic representation of different types of moraine-dammed **a** and ice-dammed **b** glacial lakes. The terms reflect the classification according to Schweizer (1957). (Modified and supplemented after Clague and Evans 2000)

◻ Fig. 9.9 Laguna Cohup, Peru, in 2017 and in 1947. The v-shaped trench in the Little Ice Age moraine is still clearly visible 76 years after the catastrophe, and the lake has grown significantly larger due to glacial retreat. (Large image: Google Earth, small image: courtesy of © Hogrefe AG 1947, all rights reserved)

9

◻ Fig. 9.10 Left: The Mattmark plain with the Allalin glacier in 1954. Right: The artificially dammed Mattmark lake in 2012. (Photo: Military Air Service, from: Schweizer 1957, under CC BY 3.0, right: Google Earth)

Schweizer (1957) classifies **ice-dammed lakes** into three types (◻ Fig. 9.8): In the 'Märjelen' type, named after Lake Märjelen at the Great Aletsch Glacier, the ice of the main glacier dams the outflow of the side valley. The Märjelen lake was artificially confined in 1895 by a tunnel, which has since lost its function due to glacier recession. The second type is called "Rofen"; here the ice from the side valley dams the river in the main valley. The type locality is the Rofen ice dam in the rear Ötztal (► Fig. 3.9), which was formed by the advancing Vernagtferner and failed eight times during the Little Ice Age, sometimes causing severe devastation. Today, this type exists mainly in more heavily glaciated mountains, where there are still extensive valley glaciers and the situation often arises that they seal off side valleys. A well-known example is Lake Merzbacher in the central Tienshan, which causes GLOFs annually and releases floods that are still observed at a Chinese monitoring station 200 km away (Glazirin 2010; Mayr et al. 2014). The third type is "Gorner", where water is dammed at the confluence of two glaciers or parts of glaciers, as is the case at the Swiss Gorner Glacier.

◘ Fig. 9.11 Lower Grindelwald Glacier (debris-covered) in 2009 with lake (left) and lake after drainage (right). (Photos: Bruno Petroni)

◘ Fig. 9.12 Left: Outburst opening at the Tête Rousse glacier. Right: destroyed school building in Bionnais. (Photo left: ETH-Bibliothek Zürich, Bildarchiv/Fotograf: Tairraz, Joseph/Hs_1458-GK-BF02–1 892–0 5/Public Domain Mark, photo right: source: Wikipedia, public domain)

Lakes that lie entirely on glaciers are called **supraglacial lakes.** At Unterer Grindelwald glacier, such a lake formed as a result of a rockfall onto the glacier surface in 2006 (◘ Fig. 9.11) and was drained by floods every year. In 2009, its volume was 1.7 million m³, more than double that of the previous year (Glaciers online 2020). Since 2010, it has been drained in a controlled manner through a rock tunnel over 2 km long.

The rarest, or at least the most difficult to detect, glacial lakes are those located in cavities in the ice and are called **water pockets.** On the Tête Rousse glacier in the Mont Blanc region, which covers only 8 ha, a water pocket burst in 1892 released 200,000 m³ of water and ice, entraining 800,000 m³ of rock material (Vincent et al. 2010) and destroying the spa town of St. Gervais (◘ Fig. 9.12), killing 175 people.

In 2008, radar measurements at Tête Rousse again showed irregularities, which were identified a year later as a 65,000 m³ accumulation of water. After the establishment of an early warning system and an evacuation plan, the water pocket was drilled with enormous technical effort and a large part of the water was pumped out. By 2013, after repeated pumping operations in 2011 and 2012, the cavity had reduced in size due to deformation of the ice to such an extent that no major water pocket can be created for the time being (Vincent et al. 2015).

9.2.2 **Breakout Mechanisms**

Moraine-dammed lakes rarely break out due to the sudden collapse of an undisturbed moraine. However, a risk factor is ice cores within the moraine that have been detached by the glacier and can lead to the destabilization of the moraine. The far more common breakout mechanism is **progressive erosion**. Each moraine is overflown by the glacial stream at its lowest point. The elevation of the outflow regulates the lake level. Moraine material is generally unsorted and has a wide grain size spectrum, ranging from fine clay to large boulders. The glacial stream flushes out the fine material, forming a boulder pavement along its bed that eventually prevents the stream from deepening further. However, during particularly high flows, this cobble layer may rupture and the stream may erode efficiently into the underlying mixed-grained substrate. Any deepening of the outlet, however, further increases the discharge from the lake, resulting in a self-reinforcing process known as progressive erosion. In this way, a deep breach can develop in the moraine in a short period of time and the lake can run dry in the form of an outburst within a day or a few days. The increased runoff that sets the process in motion may itself be triggered by an intense melting event. More common, however, are waves that are themselves caused by ice avalanches or calving into the lake. In Bhutan, 44% of GLOFs were triggered by ice avalanches and 33% by calving of large icebergs (Komori et al. 2012).

Ice dams can also break due to water pressure or earthquakes. More commonly, however, they begin to float by the increasing hydrostatic pressure of the water and the dammed water gains access to the glacier's englacial drainage system. This channel system is efficiently widened by frictional heat and the relatively warm lake water, which in turn increases flow and results in a self-reinforcing effect similar to that of progressive erosion. A special case is the situation in Iceland, where subglacial lakes are formed by volcanic activity and the eruptions are called **jökulhlaups** (Nye 1976; Björnsson 2002).

In the case of supraglacial lakes, eruption often takes place via opening crevasses; water pockets are still largely unexplored due to their rarity and inaccessibility. In the case of the Tête Rousse glacier, there were probably two cavities. In the case of the upper one, the ice roof collapsed, which increased the water pressure on the lower one, which was connected to the upper one via a channel, to such an extent that the outlet opening was literally blown out (Vincent et al. 2010).

9.2.3 **Risk Management**

One advantage in monitoring glacial lakes is that they (with the exception of water pockets) can be easily detected and monitored using optical methods (satellite images, automatic cameras). In the case of ice-dammed lakes, the floating of the ice dam that precedes an outburst can also be monitored, for example with GPS stations. Technical options for risk reduction are widespread, often based on artificial drainage (e.g. Laguna Llaca, Peru, or Tscho Rolpa, Nepal) or, less frequently, on dam raising (e.g. Grubengletscher, Switzerland). Where the risk cannot be mitigated, it is still possible to warn the population, e.g. via sirens. Modelling of flood waves from glacial lake outburst floods requires the combination of a dam-breach model (Westoby et al. 2014) or a model for the widening of an englacial channel (Petrakov et al. 2012) with a hydrodynamic model and allows the estimation of event magnitudes and the identification of areas at risk (Hagg et al. 2021).

References

Alean J (1985) Ice avalanches: some empirical information about their formation and reach. J Glaciol 31(109):324–333

Björnsson H (2002) Subglacial lakes and jökulhlaups in Iceland. Glob Planet Chang 35:255–271

Clague JJ, Evans SC (2000) A review of catastrophic drainage of moraine-dammed lakes in British Columbia. Quat Sci Rev 19:1763–1783

Dematteis N, Giordan D, Allasia P (2018) Potential precursors of ice failures in the Planpincieux glacier. 20th EGU General Assembly, EGU2018, Proceedings from the conference held 4–13 April, 2018 in Vienna, S 7543

Evans S, Bishop N, Smoll L, Murillo P, Delaney K, Oliver-Smith A (2009a) A re-examination of the mechanism and human impact of catastrophic mass flows originating on Nevado Huascarán, Cordillera Blanca, Peru in 1962 and 1970. Eng Geol 108:96–118. https://doi.org/10.1016/j.enggeo.2009.06.020

Evans SG, Tutubalina OV, Drobyshef VN, Chernomorets SS, Dougall S, Petrakov DA, Hungr O (2009b) Catastrophic detachment and high-velocity long-runout flow of Kolka Glacier, Caucasus Mountains, Russia in 2002. Geomorphology 105:314–321

Felgentreff C, Glade T (2008) Naturrisiken – Sozialkatastrophen: zum Geleit. In: Felgentreff C, Glade T (eds) Naturrisiken und Sozialkatastrophen. Springer, Berlin/Heidelberg

Flotron A (1977) Movement studies on hanging glaciers in relation with an ice avalanche. J Glaciol 19(81):671–672

Giordan D, Dematteis N, Allasia P, Motta E (2020) Classification and kinematics of the Planpincieux Glacier break-offs using photographic time-lapse analysis. J Glaciol 66(256):188–202. https://doi.org/10.1017/jog.2019.99

Glaciers online (2020) Der Gletschersee 2008 und 2009. https://www.swisseduc.ch/glaciers/alps/unterer-grindelwaldgletscher/gletschersee_2009-de.html. Accessed on 01.04.2020

Glariskalp-Alcotra (2020) Statistics about glaciers, events, victims. http://www.nimbus.it/GlaRiskAlp/GlaRiskAlpMain.asp. Accessed on 01.04.2020

Glazirin GE (2010) A century of investigations on outbursts of the ice-dammed lake Merzbacher (Central Tien Shan). Aust J Earth Sci 103(2):171–179

Haefeli R (1966) Note sur la classification, le méchanisme et le contrôle des avalanches de glace et des crues glaciaires extraordinaires. IAHS Publ 69:316–325

Hagg W, Ram S, Klaus A, Aschauer S, Babernits S, Brand D, Guggemoos P, Pappas T (2021) Hazard assessment for a Glacier lake outburst flood in the Mo Chu River basin. Bhutan Appl Sci 11:9463. https://doi.org/10.3390/app11209463

Heim A (1895) Die Gletscherlawine an der Altels am 11. September 1895. Neujahrsblatt der Zürcherischen Naturforschenden Gesellschaft auf das Jahr 1896. Zürcher und Furrer, Zürich

Komori J, Koike T, Yamanokuchi T, Tshering P (2012) Glacial lake outburst events in the Bhutan Himalayas. Glob Environ Res 16:59–70

Kotlyakov VM, Rototaeva OV, Nosenko GA (2004) The September 2002 Kolka glacier catastrophe in North Ossetia, Russian Federation: evidence and analysis. Mt Res Dev 24(1):78–83

Mayr E, Juen M, Mayer C, Usubaliev R, Hagg W (2014) Modeling runoff from the Inylchek Glaciers and filling of ice-dammed lake Merzbacher, Central Tian Shan. Geogr Ann A 96:609–625. https://doi.org/10.1111/Geoa.12061

Mool PK, Wangda D, Bajracharya SR (2001) Inventory of glaciers, glacial lakes and glacial lake outburst floods: monitoring and early warning systems in the Hindu Kush-Himalayan Region, Bhutan. ICIMOD, Kathmandu

Nagai H, Ukita J, Narama C, Fujita K, Sakai A, Tadono T, Yamanokuchi T, Tomiyama N (2017) Evaluating the scale and potential of GLOF in the Bhutan Himalayas using a satellite-based integral glacier-glacial lake inventory. Geosciences 7:77. https://doi.org/10.3390/geosciences7030077

Nye JF (1976) Water flow in glaciers: jökulhlaups, tunnels and veins. J Glaciol 17(76):181–207

Petrakov DA, Tutubalina OV, Aleinikov AA, Chernomorets SS, Evans SG, Kidyaeva VM, Krylenko IN, Norin SV, Shakmina MS, Seynove IB (2012) Monitoring of Bashkara glacier lakes (Central Caucasus, Russia) and modelling of their potential outburst. Nat Hazards 61:1293–1316

Pralong A, Funk M (2006) On the instability of hanging glaciers. J Glaciol 52:31–48

Raymond M, Wegmann M, Funk M (2003) Inventar gefährlicher Gletscher in der Schweiz, Mitteilung der VAW Nr. 182

Röthlisberger H (1981) Eislawinen und Ausbrüche von Gletscherseen. In: Kasser P (Hrsg) Gletscher und Klima – glaciers et climat, Jahrbuch der Schweizerischen Naturforschenden Gesellschaft, wissenschaftlicher Teil 1978. Birkhäuser Verlag, Basel/Boston/Stuttgart, S 170–212

Schweizer W (1957) Gletscherseen. Geog Helv 12(2):81–87

Vincent C, Garambois S, Thibert E, Lefebvre E, Meur L, Six D (2010) Origin of the outburst flood from Glacier de Tête Rousse in 1892 (Mont Blanc area, France). J Glaciol 56(113):688–698. https://doi.org/10.3189/002214310793146188

Vincent C, Thibert E, Gagliardini O, Legchenko A, Gilbert A, Garambois S, Condom T, Baltassat JM, Girard JF (2015) Mechanisms of subglacial cavity filling in Tête Rousse glacier. J Glaciol 228:61. https://doi.org/10.3189/2015JoG14J238

Westoby MJ, Glasser NF, Hambrey MJ, Brasington J, Reynolds JM, Hassan MAAM (2014) Reconstructing historic glacial lake outburst floods through numerical modelling and geomorphological assessment: extreme events in the Himalaya. Earth Surf Process Landf 39:1675–1692. https://doi.org/10.1002/esp.3617

9

Glacial Erosion

Contents

© The Author(s), under exclusive license to Springer-Verlag
GmbH, DE, part of Springer Nature 2022
W. Hagg, *Glaciology and Glacial Geomorphology*,
https://doi.org/10.1007/978-3-662-64714-1_10

> **Overview**
> Glaciers move and can erode solid and loose rocks, creating characteristic land-forms. The rate of erosion is highly variable and difficult to determine. The spectrum of resulting forms ranges from small glacial scars to the reshaping of entire valleys. Flowing water also creates special forms under the glacier, in the formation of which the pressure of the overlying ice may play a role.

The movement of a medium over the earth's surface results in a lowering effect, which is referred to as erosion in the geosciences. Depending on the medium, a distinction is made between, for example, wind erosion, fluvial erosion or glacial. The latter process does not always take place in the same way, but can differ depending on the type of movement, the type of glacier and the nature of the glacier bed. In addition to the processes of glacial erosion, the landforms that result from the erosive action of the glaciers are also shown.

10.1 Erosion Processes in Solid Rock

On smooth and intact solid rock surfaces, glacial ice, which is close to the pressure melting point and therefore well deformable, cannot have an erosive effect. Ice only reaches its maximum hardness at −70 °C and even this would only be sufficient to attack relatively soft rock. This apparent contradiction to the view of many scientists that the glaciers of the Ice Age reshaped the relief of the Alps led to a long dispute among nineteenth century geomorphologists about the effectiveness of glacial erosion, in the course of which Albert Heim (1919) was carried away to the beautiful quotation *"mit Butter hobelt man nicht* (you don't plane with butter)". The question was whether valley formation in the Alps was mainly due to glaciers or flowing water. From today's point of view, at least one aspect of the erosion process is quite comparable: Just as water alone cannot erode bare rock, neither can glacial ice. In both cases, rocks act as "cutting tools". In rivers, they are transported as boulders, and in glaciers they are stuck in the basal layers. This so-called **abrasion** and is subdivided into two sub-processes. **Polishing** refers to the polishing effect developed by the removal of the smallest protuberances, which leads to the smoothing of rock surfaces. This process dominates mainly when the eroding debris has the same hardness as the bedrock. **Striation**, on the other hand, occurs when harder particles slide over softer bedrock, causing clearly visible scratches on the rock surface. Prerequisites for abrasion to occur are the presence of basal debris and the occurrence of basal sliding. In other words, the glacier must slide over its bed and there must be rocks sticking out its bottom.

10

To date, however, there is uncertainty about the quantitative significance of the physical forces (▶ Excursus 10.1), but it is evident that both the normal pressure of the ice plays a role and the basal melt rates that control the movement of the particles against the glacier bed.

❯ Ice alone cannot attack solid bedrock, but only the rocks that are stuck in the basal ice. The glacier can only develop an abrasive effect if it glides over the bedrock and if the eroding rock particles "grow out" of the glacier bed and renew themselves again and again. Both require basal melting, which is why abrasion only occurs on temperate glaciers.

In addition to abrasion, there is a second process by which glaciers can erode bedrock. **Plucking** refers to the removal of fragments from highly fractured rock sections. This is made possible by the particles freezing to the ice or being flowed around by the ice during periods of slow ice movement and being dragged along as the speed increases again. The disruption that must precede the process is caused mainly by large pressure differences, and these occur preferentially at rock obstacles. If the flow velocity is high enough, the ice on the back side of the obstacle (lee side) cannot deform fast enough and immediately envelope the obstacle again, creating a subglacial lee-side cavity (◘ Fig. 10.1), in which atmospheric pressure usually prevails. Since the pressure of the moving ice is particularly high at the stoss-side of the rock obstruction, enormous pressure differences occur at the apex, which promote fracturing and crushing of the bedrock. Due to the pressure dependence of the melting point, refreezing occurs preferentially on the lee side due to the pressure drop, which favours the freezing and removal of loosened rock fragments.

❯ Strong pressure differences occur at bed hummocks under the ice, causing stress fields and fractures at the top and on the back. If the crushed rock fragments become surrounded by the ice or freeze to the bed and are entrained by the glacier, this is called plucking.

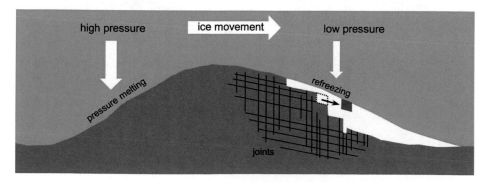

◘ **Fig. 10.1** Plucking in lee-side cavity. (After Winkler 2009, modified and supplemented)

Excursus 10.1: Abrasion Models

Since the 1970s, two theoretical models have existed which describe abrasion. According to Boulton's (1979) model, friction is decisive and erosion rates are determined by the normal pressure of the ice. Thus, ice thickness is the decisive controlling factor (■ Fig. 10.2).

Initially, the erosion rate increases with increasing ice thickness due to increasing friction. The velocity of the eroding particle is always lower than that of the ice due to frictional drag. This deceleration increases with normal pressure. Therefore, above a critical ice thickness, the erosion rate decreases again and at some point the pressure is so great that the particle velocity reaches zero. At this point, erosion turns into accumulation. In Hallett's (1979) model, it is not friction but basal melting that is decisive.

This process in fact transports rock particles that are in the ice towards the glacier bed (■ Fig. 10.3). This ensures a constant renewal of abrasive material, which would otherwise be sanded off very quickly and thus ineffective. According to this model, the abrasion rate increases with the debris concentration and the basal melting rate. In contrast to the model of Boulton (1979), this model is only applicable to debris-poor ice.

In the 1990s, the sandpaper friction model was developed specifically for debris-rich ice (Schweizer and Iken 1992). Compared to the Boulton model, it takes into account the area of the glacier bed that is not directly in contact with particles, which makes the friction force somewhat lower.

10

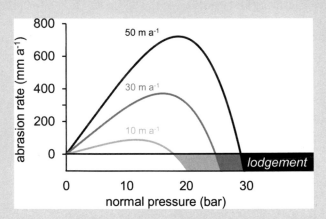

■ **Fig. 10.2** Theoretical abrasion rate as a function of normal pressure (ice overburden pressure minus water pressure), calculated for three different flow velocities using the approach of Boulton (1982). The parameters used are for a glacial bed of basalt (hardness 6–7) and debris properties as observed on a glacier in Iceland

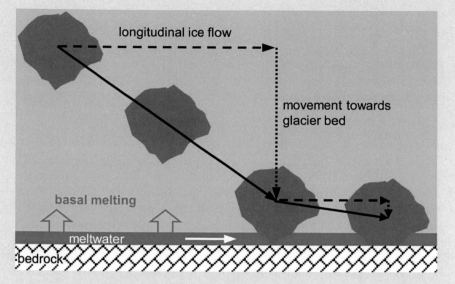

◻ Fig. 10.3 Relative motion of abrasive particles against the glacier bed. (Modified after Sugden and John 1976)

10.2 Erosion Processes in Unconsolidated Rocks

Loose rock can be subject to glacial erosion in three ways: by **pushing** at the glacier front, by **net-adfreezing** of debris, or by **thrusting** of larger frozen floes.

Thrusting at the glacier front can only exist with advancing glaciers that push unfrozen loose material in front of them like a bulldozer. Thrusting occurs when glaciers advance into frozen forelands. Thrusting causes the permafrost to constrict and deform, which is summarized as glacitectonics and the results of which are treated in more detail when forms of glacial accumulation are discussed. The prerequisite for net-adfreezing is that unfrozen sediment cools below freezing at the ice-till interface. This may be the case during the transition from warm-basal to cold-basal conditions, which can occur in polythermal glaciers whose tongues are often composed of cold ice (▶ Chap. 5). However, it is also possible that cold air intrudes beneath temperate glaciers in winter, causing the loose material at the glacier bed to freeze. Also in lee-side cavities, as shown above, there is a tendency to refreeze due to the drop in effective normal pressure. Here, not only fragments of the rock fragments freeze by plucking, but also other basal debris; this is attributed to the erosion of loose rock.

10.3 **Erosion Rates**

In general, it must be noted that parent bedrock can be eroded more effectively the more its structure is loosened by rock failure. In this context, weathering (▶ Excursus 10.2) is important, as it prepares the bedrock for erosion by glacial ice.

Direct observation of erosion rates is possible in exceptional cases where subglacial cavities and tunnels are repeatedly or permanently accessible to humans. Values between 4.5 mm per year (Upper Grindelwald Glacier; Vivian 1997) and 36 mm per year (Argentière Glacier; Boulton 1974) are reported from the Alps. These figures are difficult to compare because they depend on the flow velocity and rock type of the glacier bed. In addition, this approach does not take into account erosion by plucking. The sediment load in glacial streams, i.e. the total mass of transported suspended sediment and debris, is an indirect indication of the erosion in the catchment. Erosion rates derived in this way vary by several orders of magnitude from 0.01 mm per year on polar glaciers to 100 mm per year on fast and temperate valley glaciers (Hallet et al. 1996).

The low erosive power of polar glaciers can be explained by the type of movement. In cold glaciers without a meltwater film at the base, ice motion due to internal deformation approaches zero at the base (▶ Fig. 3.6). With the absence of basal sliding, abrasion can occur neither. Plucking by re-freezing is also ruled out when cold conditions prevail. Only when ice surrounds strongly fractured rock, plucking can exist in polar regions, but this type of erosion is comparatively inefficient. It follows that the cold-based glaciers of the polar regions are erosively quasi ineffective.

10

> **Excursus 10.2: Weathering**
>
> All processes that lead to the mechanical disintegration or chemical decomposition of solid rock are referred to as weathering. In the high mountains, there are two processes in particular that break down rock. Rocks such as granites or gneisses are formed several kilometres deep in the earth's crust under very high pressures. When the rock then comes to the surface as a result of the erosion of the overlying layers in the course of the earth's history, it already undergoes a great deal of pressure release as a result and fractures parallel to the rock surface are formed. Furthermore, frost weathering supports rock deterioration even before glacierization: water penetrates into the smallest joints and faults and expands with high pressures during nightly or winter freezing, gradually widening the rock fracture. This so-called frost wedging and the growth of ice lenses by segregation are the most productive processes of weathering in the high mountains, the result being sharp-edged frost debris. Decisive for the efficiency is the frequency of freeze-thaw cycles, which can lead to small-scale, orientation-related differences (◘ Fig. 10.4).

■ Fig. 10.4 While these southeast-facing glaciers in Tibet (1–3) have heavily debris-covered tongues, the northwest-facing ones (4 and 5), whose rock surroundings are in the shade, have much less debris cover. This is due to the more frequent freeze-thaw cycles on the south side in this very high mountain range with the summit of Kula Kangri (KK, 7538 m a.s.l.) on the disputed border with Bhutan (red line). (Google Earth)

10.4 Landforms of Glacial Erosion

Rock sections polished smooth by *polishing are* called **rock polish**, the visible scratches caused by *striation are* called **striae** (singular: striation)bedload (■ Fig. 10.5).

Striations are of great importance as evidence of the former glaciation of a site and in many places made it possible to reconstruct the extent of Pleistocene glacia-

◘ Fig. 10.5   Left: Glacial striations in the Ötztal (left) and on a boulder from the Alpine Foreland. (Photos: W. Hagg; right: Geological Collection of the University of Applied Sciences Munich)

◘ Fig. 10.6   Elongated rat tails on the valleyward side of resistant quartz knobs (dashed white lines). The ice flowed away from the observer. Sentiero glaciologico Luigi Marson, Valtellina, Italy. (Photo: W. Hagg)

10

tion. Furthermore, the scratches even indicate the flow direction of the ice. In the case of alpine glaciers this is rather obvious as it is largely determined by the relief, but in the case of former ice sheets it is often not so easy to deduce and has sometimes changed several times due to the shifting of the ice centres during a glacial period. Here, glacial scars can provide important clues to the course of the glaciation history. They often widen in the direction of flow by blunting of the scoring particle. This can occur slowly (wedge striae) or abruptly (nailhead striae) (Benn and Evans 1998) and prevents 180° misinterpretation of flow direction. Scrapes may exhibit a sudden lateral offset due to rotation of the particle.

Small-scale inhomogeneities in rock hardness can be modeled out of the glacier bed by abrasion. Local occurrences of harder rock are less eroded and grow as a knoll from their surroundings. Because erosion is reduced on the lee side of the rock knob, an elongated, streamlined tail forms here. Such shapes are called **rat tails**; they exist at several scales, from millimeters to meters. In ◘ Fig. 10.6 one can see three rat tails of medium size, which have formed in the protection of decimetre-sized, more resistant quartz cores.

> For effective abrasion, the hardness of the eroding particles is decisive. If they are as hard as the bedrock, the latter is finely polished; if they are harder, it is scratched (glacial abrasion). If the subsurface has individual harder spots, these become knolls or nodules with elongated tails (rat tails) on the lee side.

Other small-scale forms of erosion are **friction cracks**; they occur during jerky ice movement when a single particle exerts high punctual pressure on the bedrock and parts of it are chipped out. This results in crescentic gouges and fractures in the rock, the open or concave side of which usually faces up-ice. Fine cracks are called *parabelrisse* (singular: *parabelriss*), larger ones *sichelbrüche* (singular: *sichelbruch*). The rarer counterparts with open sides facing down-ice are called **lunate fractures**. **Chattermarks** are transitional forms between striae and parabolic cracks; they are formed by jerking stick-slip motion and are only a few centimetres wide (◻ Fig. 10.7).

◻ **Fig. 10.7** **a** Lunate fracture shortly after formation. **b** Crescentic gouge. **c** Chattermarks. Ice flow was from right to left in each case. (Photos: **a** Archive Geodesy and Glaciology; **b** and **c** W. Hagg)

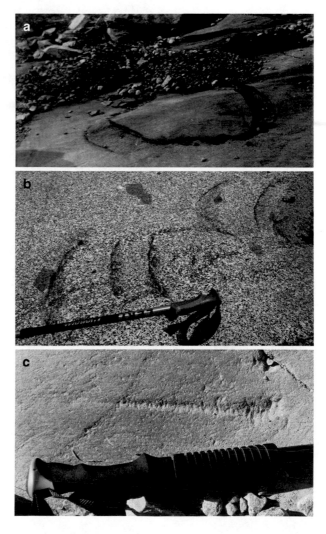

As a special case of glacial erosion, the subglacial, glaciofluvial forms will be treated here. Some are indistinguishable from other forms shaped by flowing water, while others are formed exclusively under glaciers. In the latter, the high water pressure due to the weight of the glacier seems to play a part. There is a great variety of these centimetre to metre sized formations, they are subsumed in the literature as **P-forms (plastically moulded forms).** Originally they were thought to be abrasive forms (Dahl 1965), but it is now considered certain that liquid water is at least involved in their formation. They are divided into longitudinal, transverse and non-directional forms according to their orientation to the flow direction of the ice (Kor et al. 1991).

Muschelbrüche (singular: *muschelbruch*) and *sichelwannen* (singular: *sichelwanne*) are crescent-shaped structures and belong to the transverse forms. They look similar to *sichelbrüche*, but have softer shapes and can become larger. Round longitudinal channels are called **furrows** and can be elongated or winding (◻ Fig. 10.8).

The best-known of the non-directional P-forms are **potholes**, which often origin in stationary crevasses where water from a great height hits the bedrock and the swirling eddies carve out round depressions with the help of entrained rock particles (◻ Fig. 10.9).

Larger rock obstacles in the glacier bed are reshaped into *roches moutonnées* (singular: *roche moutonnée*) by the combined effect of abrasion and plucking. On the stoss side, increased pressure melting favours basal sliding. This leads to strong abrasion and the formation of flat up-ice sides, on which striations and friction cracks often indicate the high overburden pressure. The lee side is steeper and more jointed due to the quarrying and plucking that occurs here, giving the bedrock bumps their typical asymmetrical shape (◻ Fig. 10.10). Normally, they are metres to tens of metres in size and are longer than they are high; sometimes even larger forms can be referred to as *roche moutonnées*.

Shapes related to *roche moutonnées* are **rock drumlins**. They are formed at lower ice flow velocities, resulting in less pressure on the stoss face and no cavity

◻ **Fig. 10.8**   Longitudinal P-form at Blaueis, the northernmost glacier in the Alps. (Photos: W. Hagg)

🔲 **Fig. 10.9** Potholes at the Maloja Pass. (Photo: W. Hagg)

🔲 **Fig. 10.10** *Roches moutonnées* of various sizes. **a** In the urban area of Helsinki. **b** In Valchia-venna. **c** Mittlerer Burgstall in the Hohe Tauern. (Photos: W. Hagg)

on the lee face. Due to the uneffective abrasion and the lack of plucking they have a steep front and a flat back. If the pressure on the stoss side and thus the abrasion increases somewhat without creating a subglacial cavity, symmetrically shaped **whalebacks** can develop.

When, outside of mountains, large, isolated rock hills are deformed by glaciers, so-called **crag-and-tails** are formed. These are similar in origin to rat tails, but with a length of tens of metres to kilometres they have a clearly different spatial dimension. Here, too, the resistant rock produces a pressure shadow on its lee-ward side, reducing the erosive power and creating an elongated "tail". These hills

Fig. 10.11   City panorama of Edinburgh with two prominent crag-and-tails. (Photos: W. Hagg)

dominate, for example, the cityscape of Edinburgh (Evans and Hansom 1996), where the castle is perched on a hard volcanic plug (crag) and the main shopping street extends down the flat leeward side (tail) (Fig. 10.11). In Germany, the Hohentwiel near Singen can be attributed to this group of landforms.

> A *roche moutonnée* is the normal shape formed from a rock that is overflowed by ice. Increased abrasion flattens the stoss side and plucking in the subglacial cavity steepens the lee side.

If a subglacial cavity forms on the lee side in which basal debris is accumulated, the tail may also consist of glacial deposits. In this case, crag-and-tails are not pure erosional forms, but mixed forms of selective erosion and sedimentation.

As already indicated in ▶ Sect. 2.1, glaciers can only form if flattening zones or concave terrain areas exist above the climatic snow line where sufficient snow can accumulate. Such situations are often found at the base of walls, where a steep rock face meets a flatter talus slope. Avalanche snow accumulates here and remains for a particularly long time or even lasts the entire summer. From such patches of snow, processes such as snow creep and sliding, as well as increased chemical weathering due to constant moisture supply, create larger flat shapes called **nivation niches.** The larger these niches become, the more snow can accumulate there, eventually producing moving glacial ice via metamorphism of the snow. As small glaciers rotate in such depressions, glacial erosion occurs. A semicircular armchair hollow formed in this way is called a **cirque (corrie)**, both when it is still filled with ice and after the glacier has melted (Fig. 10.12). Cirques have a steep headwall, a glacially eroded and sometimes overdeepened floor, and a dam at the outlet where the erosional action subsides. When such a bowl-formed hollow fills with water, a lake **(tarn)** is formed. By incision of the outflow into the dam, these often have a limited life and may dry up again. Mountains with cirques on several sides, which intersect to form ridges (so-called **arêtes**) and whose summits are therefore sharpened like pyramids, are called **horns**.

> Cirques are typical landforms of glacierized mountains and often serve as evidence of former glaciation.

If glaciers grow beyond their cirque, they flow into the valley below and thus become valley glaciers. In doing so, they typically reshape the cross section of the

◘ Fig. 10.12 Glacial cirques in the Aosta Valley. (Photos: W. Hagg)

valley. The linear vertical erosion of flowing water usually creates more or less V-shaped valleys. Under glaciers, the main subglacial drainage channel flows in the valley center, and its water pressure counteracts the ice overburden pressure. For this reason, the highest flow velocities and thus erosion rates are not at the deepest part of the cross section, but somewhat up the valley walls (Harbor 1992). This causes the valley floor to widen and erosion on the valley slopes causes them to steepen. Eventually, a U-shaped equilibrium profile is formed in this manner, which continues to deepen as a whole. Especially in erosion-resistant rocks, the ideal form of a U-shaped valley or **trough** may be formed (◘ Fig. 10.13). The action of glaciers thus reshapes V-shaped valleys into troughs. However, one should be aware that the valley shape depends on other, mainly geological factors. For this reason, glacially formed valleys often deviate from this ideal form and are asymmetrical or parabolic in shape (Baumhauer and Winkler 2014). Similarly, U-shapes may not reflect the actual shape of the valley in bedrock, but are merely faked to the observer by concave sedimentary fillings. Drowned troughs that end in the sea are called **fjords** after they become ice-free.

> ❯ Troughs are considered to be the ideal typical large-scale form of a glacial mountain relief. However, not every glaciated valley develops an ideal trough shape and under certain circumstances unglaciated valleys can also form U-shaped profiles.

Glacier erosion affects not only the cross section of valleys, but also their longitudinal profile. Since abrasion and plucking depend on ice thickness, velocity and rock hardness, vertical erosion varies over the course of a long valley glacier. Unlike water, glacial ice can also flow uphill, resulting in deeper eroded basins, e.g. in valley sections with softer bedrock and intervening elevated hard rock bars called **riegel**. The term overdeepening, which appears in this context, is unfortunately not used uniformly. Sometimes it means the depth below the present valley floor, which is often composed of postglacially accumulated fluvial gravels (a in ◘ Fig. 10.14) and sometimes it refers to the elevation of a down-valley sill (b in ◘ Fig. 10.14). For fjords, overdeepening often refers to erosion below present-day sea level.

◘ Fig. 10.13 Two valleys in the Georgian Caucasus. Top: Trough. Bottom: While the rear part has been transformed into a U-shaped valley, the front part was obviously not glaciated long enough and has retained its V-shaped cross section. (Photos: W. Hagg)

10

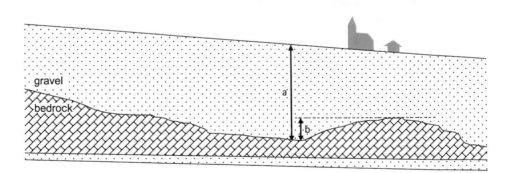

◘ Fig. 10.14 Two ways of defining glacial overdeepening

As the erosion rate depends on glacier discharge, it increases abruptly at the junction of two glaciers (**confluence**), creating a step in the longitudinal profile. Where a small glacier of a side valley joins the main valley, the difference in erosion power is particularly large. The downward erosion of the side valley cannot keep pace with that of the main valley, and step forms between the two valley

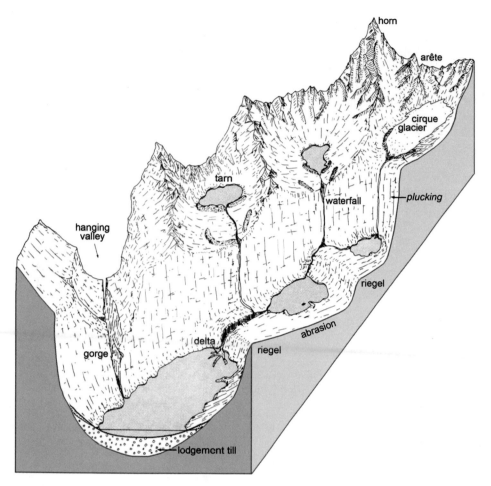

◻ Fig. 10.15 Large-scale landforms of glacial erosion. (From: Selby 1985, courtesy of © Oxford Publishing Limited, slightly modified reprint by permission of PLSclear)

floors (◻ Fig. 10.15). After deglaciation, a **hanging valley** is formed in this way, rivers often overcomes the difference in elevation as a waterfall or in a steep gorge. When glaciers flow apart (**diffluence**), the erosion rate decreases accordingly and sills (bars) form in the longitudinal profile.

❯ The longitudinal profile of glacial valleys considers the bedrock that has been shaped by glacial activity, rather than the present-day profile of the valley floor if it has been subsequently filled with river gravels.

At the end of the last glacial period, the large Alpine valleys such as the Inn valley must be imagined as a chain of lakes that filled the overdeepened basins. Due to the strong sediment input, they quickly silted up and the valley floors were filled with sediment, which is often several 100 m thick (Preusser et al. 2010).

Outside the Alps, piedmont glaciers have created large depressions through glacial erosion of the loose material and deposition at their former margins. These so-called *Zungenbecken* (tongue basins) and often fill with water and form lakes. When a large river entered such a lake, it silted up again already in the Late Glacial. This happened, for example, at Lake Rosenheim, which was more than five times the size of today's Chiemsee, but disappeared again within a few thousand years due to the strong sediment load of the Inn River and the lowering of the outlet. The lakes without large tributaries, such as Lake Starnberg, are almost preserved in their original size.

References

Baumhauer R, Winkler S (2014) Glazialgeomorphologie – Formung der Landoberfläche durch Gletscher. Bornträger, Stuttgart

Benn DI, Evans DJA (1998) Glaciers & Glaciation. Hodder Arnold, London

Boulton GS (1974) Processes and patterns of glacial erosion. In: Coates DR (ed) Glacial geomorphology. State University of New York, Binghamton, pp 41–87

Boulton GS (1979) Processes of glacier erosion on different substrata. J Glaciol 23:15–38

Boulton GS (1982) Processes and patterns of glacial erosion. In: Coates DR (ed) Glacial geomorphology. Springer, Dordrecht, pp 41–87

Dahl R (1965) Plastically sculptured detail forms on rock surfaces in northern Nordland, Norway. Geogr Ann 47:83–140

Evans DJ, Hansom JD (1996) The Edinburgh Castle crag-and-tail. Scot Geogr Mag 112:129–131

Hallet B, Hunter L, Bogen J (1996) Rates of erosion and sediment evacuation by glaciers: a review of field data and their implications. Glob Planet Chang 12:213–235. https://doi.org/10.1016/0921-8181(95)00021-6

Hallett B (1979) A theoretical model of glacier abrasion. J Glaciol 23:39–50

Harbor JM (1992) Numerical modeling of the development of U-shaped valleys by glacial erosion. Geol Soc Am Bull 104:1364–1375

Heim A (1919) Geologie der Schweiz, Bd 1. Tauchnitz, Leipzig

Kor PSG, Shaw J, Sharpe DR (1991) Erosion of bedrock by subglacial meltwater, Georgian Bay, Ontario: a regional view. Can J Earth Sci 28:623–642

Preusser F, Reitner J, Schlüchter C (2010) Distribution, geometry, age and origin of overdeepened valleys and basins in the Alps and their foreland. Swiss J Geosci 103:407–427

Schweizer J, Iken A (1992) The role of bed separation and friction in sliding over an undeformable bed. J Glaciol 38:77–92

Selby MJ (1985) Earth's changing surface: an introduction to geomorphology. Oxford University Press, Oxford

Sugden DE, John BS (1976) Glaciers and landscape. Edward Arnold, London

Vivian R (1997) La mesure de l'érosion des glaciers tempérés; essai de synthèse. Rev Géogr Alp 85:9–32

Winkler (2009) Gletscher und ihre Landschaften. WBG, Darmstadt

10

Glacial Sedimentation

Contents

> **Overview**
> Moraines are landforms of glacial deposition. They are formed in different locations and by different, active and passive processes. For this reason, they can be classified according to the type of formation and the place of origin. In addition, there are special forms of glacial sedimentation that cannot be pressed into the classical scheme of moraines. A literally flowing transition often exists from pure glacial deposits to those of glacial meltwater, which also forms typical topographies or even entire landscapes. A regularity in the spatial sequence of glacial landforms is offered by the concept of the "glacial series", which was formulated more than 140 years ago.

After rock particles have been eroded and transported over a shorter or longer distance, they are eventually redeposited elsewhere. This process is also called sedimentation. In the case of the glacial transport system, the loose rock is deposited at the edge of the ice (frontal or lateral) or under the ice (subglacial). The resulting landform is called a **moraine**, while the material is referred to as **till**.

11.1 Processes of Glacial Accumulation

Active accumulation under moving ice is called **lodgement**. The process occurs when the frictional drag is greater than the kinetic energy and the particle velocity drops to zero. In Boulton's (1974) model, there is a smooth transition from abrasion to lodgement (▶ Excursus 10.1), controlled mainly by normal pressure (ice pressure minus water pressure) and debris concentration. For Hallett (1979), abrasion and lodgement are independent processes because abrasion is controlled by basal melt rates (▶ Excursus 10.1).

Melt-out is the passive melting of very slow or stagnant ice (dead ice). This can occur subglacially when basal melting moves rock particles in the ice towards the glacier bed and sediments them there. When the entire ice thaws, also supraglacial material is eventually deposited by melt-out.

Dumping refers to the lateral sliding of supraglacial rock material at the glacier margin. It can lead to the formation of ridges, but these are to be distinguished in their formation from moraines pushed up by the glacier.

Glaciofluvial accumulation is the deposition by glacial meltwater. It occurs mostly proglacially and to a lesser extent in subglacial channels and cavities.

11.2 Till

The most important parameters for the characterization of sediments are grain size, sorting, rounding and bedding (▶ Excursus 11.1 ◘ Fig. 11.1).

Rock debris in mountains is formed mainly by frost weathering of solid rock or by landslides. The result is primarily coarse and sharp-edged, but it is altered in different ways during onward transport, depending on the medium.

11

◘ **Fig. 11.1** Examples of clasts with different rounding, from left to right: angular (Wetterstein limestone), subangular (radiolarite), subrounded (Julier granite), rounded (Reiseselsberg sandstone). (Photos: W. Hagg)

Moraine material (till) was transported by flowing ice, either on the glacier, in the glacier, under the glacier, or at the glacier terminus. The mode of transport has an impact on the exact material properties, but the following is more or less true for all glacial sediments: glaciers transport material regardless of grain size and deposit it all together. For this reason, moraine material shows a broad grain size spectrum and no sorting at all (◘ Fig. 11.2a). Because no uniform sedimentary bodies are formed in this way, no bedding occurs. During transport, the particles are moved comparatively little against each other, they are rather pushed. For this reason, the deposits show only a slight rounding (angular to subangular).

Flowing water, in contrast, can only transport certain grain sizes at certain flow velocities. The faster it flows, the greater its transport capacity, and the larger the largest boulders that can be carried along. When the flow velocity falls below a typical value for each grain size, the kinetic energy of the water is no longer sufficient for transport and the corresponding particles remain in the stream or river bed. Thus, when the flow velocity decreases, gravel is deposited first, then sand, and finally the suspended particles (silt and clay). This results in sorting, which also allows layering and bedding. Rolling transport in flowing waters promotes rounding, and the clasts are usually subrounded to rounded (◘ Fig. 11.1).

❯ Glacial deposits are unsorted, angular and unbedded. This clearly distinguishes them from fluvial deposits, for which the exact opposite is true for these sediment characteristics.

Lodgement till is formed by lodgement under moving ice. Due to the high pressure under the glacier, it is compact and the particles often have sctratches. A high proportion of fine material (**boulder clay**) is typical of lodgement till, which therefore often has waterlogging properties. However, coarser components up to boulder size may be embedded in the fine-grained matrix, in which silt is often the dominant grain size.

The **melt-out till** is formed not by settling but by melting out. In contrast to the lodgement till, the material is only deposited after melting, but no longer moved. For this reason, the melt-out moraine can only be formed in the case of dead ice or

11

◘ Fig. 11.2 Properties of different deposits. (**a**) Unsorted, angular till. (**b**) Sorted, rounded beach pebbles. (**c**) River deposits sorted by flow velocity, fine material (bottom) in still-water areas and coarser gravel in stronger flow. (**d**) Bedding created by a succession of different sedimentation conditions. (Photos: W. Hagg)

very slowly flowing ice. Melt-out can occur subglacially by basal melting processes at debris-rich glacier beds (subglacial melt-out till), but due to the comparatively low melt rates this process is not very productive. A large part is formed on the glacier (supraglacial melt-out till) and remains as debris cover on the glacier surface until final sedimentation by thawing of the ice. Both variants will be deposited under an abundance of meltwater, which causes a stronger flushing out of fine material, especially in the case of supraglacial melt-out till. For this reason, melt-out till is generally somewhat more sorted, coarser-grained (◘ Fig. 11.3), and more permeable to water than lodgement till. However, supraglacial melt-out till formed on shear planes of the ice may also be fine-grained. Even pronounced sort-

◘ Fig. 11.3 Melt-out till over lodgement till in a gravel pit near the Andechs Monastery, Southern Bavaria. (Photo: W. Hagg)

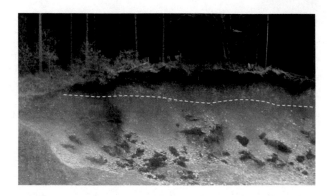

ing and bedding are possible due to the influence of meltwater. This shows how complex and interlocking the interrelationships and processes are and how difficult a stringent classification therefore is.

In addition to the main types just presented, there are a myriad of special types (Benn and Evans 1998), but these are beyond the scope of this book and should be reserved for the experts, i.e. sedimentologists.

Because the sedimentation process of particles in glacial meltwater does not differ from that in other flowing waters, the properties of glaciofluvial sediments naturally also differ accordingly from those of purely glacial ones. Because glaciofluvial sediments were only transported by flowing water for a comparatively short time and some glacial properties such as angularity or striations may still be preserved, they occupy an intermediate position between glacial and fluvial sediments.

Excursus 11.1: Morphometry of Sediments

The **grain size** describes the largest diameter of a particle; a distinction is made between the main classes boulders (>200 mm), cobble (>63–200 mm), gravel (>2–63 mm), sand (>0.063–2 mm), silt (0.002–0.06 mm) and clay (<0.002 mm).

Sorting is a measure of the dispersion of particle sizes. The better sorted a sediment is, the more uniform the size of the particles.

The most important aspect of grain shape is roundness. While in construction geology six gradations are made according to the European standard EN ISO 14688, in geomorphological coarse sediment analysis (Leser 1977) a distinction between the four rounding degrees angu-

lar, subangular, subrounded and rounded (**◘ Fig. 11.1**) is usually most appropriate because it can also be made purely visually in the field (Reichelt 1961).

In sedimentology, a stratum refers to a uniform sedimentary body. The term **bedding** refers to a succession of different layers, the change of material at the layer boundary indicates a change in depositional conditions. Sediments can not only be either bedded or unbedded, due to the recognizability of the strata, there are also intermediate stages (slightly bedded) and extreme forms (strongly bedded). Furthermore, the bedding can not only be horizontal, but also wavy, curved or lenticular.

11.3 Moraine Types

The terminology of the different types of moraines is extremely complex due to the interrelation of processes and sedimentation environments. Various attempts at a comprehensive classification according to the formation process or the place of formation can be found in the literature. Here we adopt a pragmatic mixed approach, making an initial classification according to place of deposition, followed where necessary by a finer subdivision according to the dominant process. Moraine is defined as debris that is still on or in the glacier and thus still in motion, as well as sediment that was finally deposited at the glacier margin or after the glacier has melted.

Supraglacial moraines have already been discussed in ▶ Chap. 7; these are the debris covers that reached the ablation area by rockfall or rock avalanche, or englacial debris **(inner moraine)** that was transported to the surface by emergent ice movement and becomes supraglacial melt-out till sedimentologically.

Ground moraine is subglacial and morphologically inconspicuous, it usually lies as a more or less thin layer on the subsurface and can be flat to undulating or hilly. The moraine material can be formed as lodgement till or subglacial melt-out till.

Lateral moraines are formed by dumping at lateral glacier boundaries. Since the lateral glacier margin often has the same or a similar position during different advances, it is clearly more common than in end moraines (see below) that deposits from different advances overlap. For this reason, lateral moraines are often higher and morphologically more conspicuous than end moraines (Röthlisberger and Schneebeli 1979). An example of this is shown in (◘ Fig. 11.4).

11

◘ **Fig. 11.4** White arrows show lateral moraines (top) and terminal moraine (bottom) of the Little Ice Age highstand at Ochsentalferner, Silvretta. (Photos: W. Hagg)

A special form of lateral moraines occurs with platy rock fragments. They often settle parallel to the ice surface, creating a stratification. Such layered lateral moraines (Humlum 1978) can be recognized by the fact that individual rock slabs protrude from the inner side of the moraine at the angle of the former ice surface (◘ Fig. 11.5).

Medial moraines are most often formed at the confluence of glaciers or glacial branches from the union of two lateral moraines (*ice-stream interaction type*; Eyles and Rogerson 1978). In this way they demonstrate how many firn basins or lateral valleys make up a large valley glacier (◘ Fig. 11.6).

A second possibility for the formation of medial moraines exists exclusively in the ablation area, where a relative movement in the ice to the surface (emergence) exists due to the combination of melt at the surface and the flow of the glacier. In this way, basal debris can reach the glacier surface and become supraglacial debris. In the case of large punctual debris accumulation, for example in the case of pronounced plucking at rock bumps, so-called ablation dominant moraines can develop according to this principle (Eyles and Rogerson 1978; Brook et al. 2017).

◘ **Fig. 11.5** Layered lateral moraine on Guslarferner, Ötztal. (Photos: W. Hagg)

□ Fig. 11.6   Ice stream
interaction (1) and ablation
dominant (2) type medial
moraines in the Karakoram.
(Google Earth)

After the final deposition of medial moraines following the melting of the glacier, the shape is usually no longer visible. Only if they consist of clearly different looking rock than that at the place of deposition, they can remain recognizable as a striped pattern.

End moraines (terminal moraines) are deposited at the front of the glacier. Here, too, dumping causes depositional processes, but these only have a minor quantitative effect. Without pushing or thrusting, i.e. by dumping alone, no moraine ridges would form. The reason for this is that the position of the glacier front fluctuates strongly and does not remain in place long enough. This means that there can be no terminal moraine where the glacier terminus is on bedrock, even with sufficient upper moraine. Terminal moraines are thus only formed when the glacier advances into its unfrozen or frozen foreland of loose material.

Pushing, already mentioned in the context of glacial erosion, refers to the bulldozing effect of the advancing glacier front on unfrozen unconsolidated sediments. The material is pressed up and deformed during transport. As soon as the advance ends, the material is deposited. The resulting ridge, which is often steeper on the outside than on the inside, is called a **push moraine**. During long-distance pushes, a considerable part of the material piling up at the front always gets under the glacier, which is why push moraines are never higher than a few metres (**□** Fig. 11.7). Particularly low forms result from short, seasonal winter advances: due to the omission of ablation during the cold season, small, positive changes in length of a few metres can occur during sustained ice movement, which push up small ramparts (**□** Fig. 11.7b).

In narrow mountain valleys, terminal moraines are often subsequently eroded by flowing water or covered by fluvial sediment.

When glaciers advance into frozen forelands, the glacier front is always cold basal, and no basal sliding occurs. The pressure of the glacier and the frozen basement is transferred to the permafrost in the foreland, which is shortened and behaves like solid rock: Compression and folding occur. Where unbroken deformation is no longer possible, entire floes are dislocated long failure planes (faults).

◘ Fig. 11.7 (a) Push moraines of the late nineteenth century in the Mælifell Sandur, Iceland. The glacier Sléttjökull is still visible at the right edge of the picture. (b) Small squadrons of push moraines (winter moraines) at Sléttjökull. (c) Push moraines at Hornkees, Zillertal. (Photos: W. Hagg)

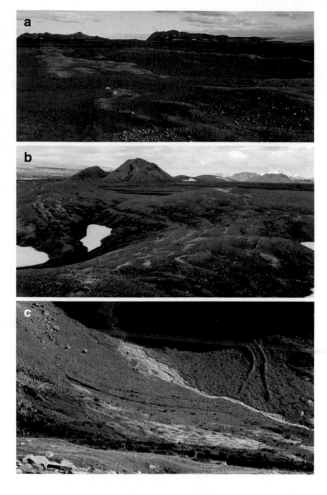

This can be described as **thrusting,** analogous to the processes during mountain formation. This constricting effect of the so-called **glacitectonics** also leads to the formation of ridges, the **thrust moraines**. They are characteristic for polar regions, but also the large Pleistocene moraine complexes in Central Europe were formed in this way. Since many layers of frozen sediment can be pushed on top of each other during this type of formation, thrust moraines can reach heights of tens of metres. In the Alpine foothills, these deposits form the large moraine ridges around the lake-filled tongue basins. A nice example of the involvement of glacial tectonics in the formation of moraines in the Alpine foreland is shown in (◘ Fig. 11.8).

❯ Terminal moraines only form when glaciers advance into unconsolidated sediments. If this is unfrozen, relatively low push moraines are formed; if the sediments are frozen, large blocks can shear off and overlap, allowing higher thrust moraines to form.

◘ **Fig. 11.8** Folding of originally horizontally deposited gravels: Evidence for glacitectonics in a gravel pit in Ostallgäu. (Photo: Hermann Jerz)

11.4 Special Forms

Flutings, fluted moraines or glacial flutes are a special form in the glacial depositional environment. These are ridges in the flow direction of the glacier, which are formed by the extrusion of ground moraine into subglacial cavities (Ives and Iverson 2019). This occurs preferentially on the lee side of rock obstructions in water-saturated, deformable substratum. Loose material is forced into the cavity by the overburden pressure of the ice. With each successive glacial movement, newly created cavity is immediately filled, so that the shape becomes longer and longer over time (◘ Fig. 11.9).

Drumlins (from the Irish Gaelic *droimnín* for "small ridge") are ice-motion-parallel, streamlined forms with steep stoss sides and shallow leeward sides. They are composed of loose material and are typically a few tens of metres high and at most a few 100 m long (◘ Fig. 11.10). At the margins of Pleistocene ice sheets and piedmont glaciers they often occur associated as "drumlin swarms". Their genesis is still controversial, and it is not even clear whether they are erosional or sedimentary forms, or a combination of both (Fowler 2018). The argument against a pure erosional form is that they preferentially occur in areas where erosional capacity is declining. This is, for example, where piedmont glaciers diverge into individual lobes. An important prerequisite for their formation seems to be a bed of deformable sediments of varying hardness over which the glacier flows. In this process, the more resistant parts are flowed around in a streamlined fashion, with pronounced erosion on the frontal side and more protection from excavation or even sedimentation occurring on the lee side. Here a certain analogy to crag-and-tails, which however always have a core of bedrock, is obvious. At the ice margin, differences in hardness may be due to sediment distribution. At the glacier gates, coarser rock debris is sedimented by the glacial stream, and this is more resistant to frontal advance than the fine sandy material in between. Other possible origins for the differences in hardness are also conceivable (Boulton 1987).

◘ Fig. 11.9 Fluted moraine in the forefield of Sléttjökull, Iceland. The arrows on the glacier terminus in the foreground point to three particularly prominent flutes. (Photo: W. Hagg)

◘ Fig. 11.10 Drumlin in the Eberfinger Drumlinfeld, the largest assemblage in Bavaria with 360 individual forms. The wooded ridge behind belongs to the Würm moraine complex around Lake Starnberg, which can also still be seen. (Photo: W. Hagg)

11.5 Glaciofluvial Landforms

Glaciofluvial sediments were deposited by glacial meltwaters. They are distinguished from pure fluvial deposits by the fact that glacial ice was either directly involved in their formation or that the form can only be explained by the huge volumes of meltwater during Pleistocene deglaciation. Their material is often subangular and, as already mentioned, represents a transitional form between the angular glacial and the rounded fluvial deposits.

The huge glaciofluvial outwash plains beyond the terminal moraines are called **sandurs**, which is the Icelandic term for "sand" and already indicates the predominant small grain sizes of the substrate (◘ Fig. 11.7a). In northern Germany, too, the corresponding sediments consist of the sands from the crystalline rocks transported in from the Scandinavian ice sheet. In southern Germany the transport routes were much shorter, which is why the deposits here consist of coarser gravels; they are called **gravel plains**. When outwash plains are attached to a single breakthrough through the terminal moraine, they form shallow alluvial fans. When the river later cuts into this deposit, the cone shape often creates trumpet-shaped valleys that widen downstream. Within the mountains, spatially limited sandurs develop due to the lack of space, they are called **valley trains** (◘ Fig. 11.11).

Fig. 11.11 Valley train below the Mandrone glacier, Adamello-Brenta Nature Park, Italy. (Photo: W. Hagg)

> Glaciofluvial sediments of fine material are called outwash plains or sandurs. In the Pleistocene, large outwash plains were formed by the Scandinavian Ice Sheet; in the Alpine foothills, the sediments consist of coarser gravels and are called gravel plains; in the mountains, spatially limited valley trains are formed.

A **kame** is a debris accumulation in crevasses or between dead ice blocks that remains as a mound after the ice has melted. Depending on the specific depositional site, different forms may develop. Kame plateaus can develop from supraglacial accumulations, kame terraces from marginal accumulations between glacier and valley slope, and so-called Moulin kames (**□** Fig. 11.12) from fillings of glacial mills.

An **esker** is an elongated embankment (**□** Fig. 11.13) formed by accumulation in the subglacial channel. The material is rounded and bedded; due to pressure flow in subglacial tubes, the ramparts can also overcome counter-slopes. Eskers may be branched (braided esker) or unbranched (single ridge esker) and are rare in the high mountains.

In places where ice is separated from the active glacier and melts passively as dead ice, the glaciofluvial sedimentation often causes these ice blocks to become buried partially or completely in the gravel. Since they keep their location from being gravelled, hollow forms remain here after melting, the **kettle holes**. When filled by rain, a **kettle pond** is formed, which normally falls dry in summer. A connection to the groundwater creates a **kettle lake** which exists all year round. When large parts of the piedmont glaciers and ice sheets were detached during the Late Glacial, the excess supply of meltwater created entire landscapes of chaotic topography (**□** Fig. 11.14) in which kettle holes are often associated with kames (kame-and-kettle topography; Benn and Evans 1998).

> When larger parts of glaciers are detached during downwasting, ice disintegration features are formed. Debris deposits on and in the ice or between ice bodies are called kames, fillings of subglacial channels are called eskers. In places where blocks of dead ice are buried by glacial outwash, kettle holes are formed after melting.

◧ Fig. 11.12 Moulin kame near the Andechs Monastery, Southern Bavaria. (Photo: W. Hagg)

◧ Fig. 11.13 The esker of Maxlrain extends over 2 km in the Alpine foothills. Only the northernmost part (foreground) is unwooded. (Photo: W. Hagg)

◧ Fig. 11.14 Evidence of the Ice Age at the Osterseen: Kames (ka) and kettles holes (ke) associate to form a so-called kame-and-kettle topography. (Photo: W. Hagg)

11.6 The Glacial Series

The spatial change of glacial erosional and depositional landforms and the transition to glaciofluvial deposits often shows a regular zonation. This was recognized early on and described by Penck and Brückner (1901–1909) as a model conception of the "glacial series". This refers to the ideal-typical sequence of landforms from the tongue basin and the ground moraine landscape with drumlin fields via the terminal moraine and sandurs to glacial meltwater valleys *(Urstromtäler,* singular: *Urstromtal)* (🔲 Fig. 11.15). The latter are valleys running parallel to the ice margin, which drained the meltwater of the Scandinavian Ice Sheet along the terminal moraine ridges. In the Alpine foothills there are no glacial meltwater valleys because the terrain here slopes away from the ice and the meltwater could therefore flow freely.

Glacial series are thus almost inevitably crossed, in the opposite direction to that just described, when travelling in southern Germany towards the Alps. A particularly beautiful variant is offered by the railway line from Munich to Starnberg, which leads through the breach of the Leutstetten end moraine by the river Würm (🔲 Fig. 11.16), which is the eponymous type locality for the last glacial period.

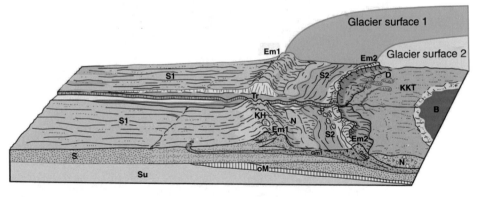

🔲 **Fig. 11.15**   A glacial series consisting of two glacial advances (glacial complexes: 1 = the older, 2 = the younger). TB tongue basin, KKT kame-and-kettle-topography, D drumlin, Gm ground moraine, Em end moraine, S sandur, KH kettle hole, E end moraine of the furthest advance, oM old moraine of a previous glaciation, Su pre-glacial substratum. (From: Rögner (2020), Glazial geprägte Landschaften, published 2020 by Springer Spektrum, slightly modified and translated reprint with permission of SNCSC)

🔲 **Fig. 11.16**   View to the south into the tongue basin of Lake Starnberg, in the foreground the ridge of the first recessional phase of the Leutstetten terminal moraine, which the Würm breaks through on the right outside the picture. (Photo: W. Hagg)

References

Benn DI, Evans DJA (1998) Glaciers & glaciation. Hodder Arnold, London

Boulton GS (1974) Processes and patterns of glacial erosion. In: Coates DR (ed) Glacial geomorphology. State University of New York, Binghamton, pp S 41–S 87

Boulton GS (1987) A theory of drumlin formation by subglacial sediment deformation. In: Menzies J, Rose J (eds) Drumlin symposium. Balkena, Rotterdam, pp S 25–S 80

Brook MS, Hagg W, Winkler S (2017) Contrasting medial moraine development at adjacent temperate, maritime glaciers: fox and Franz Josef Glaciers, South Westland, New Zealand. Geomorphology 290:58–68

Eyles N, Rogerson RJ (1978) A framework for the investigation of medial moraine formation: Austerdalsbreen, Norway, and Berendon Glacier, British Columbia, Canada. J Glaciol 20(82): 99–113

Fowler AC (2018) The philosopher in the kitchen: the role of mathematical modelling in explaining drumlin formation. GFF. https://doi.org/10.1080/11035897.2018.1444671

Hallett B (1979) A theoretical model of glacier abrasion. J Glaciol 23:39–50

Humlum O (1978) Genesis of layered lateral moraines. Geogr Tidsskr 77:65–71

Ives LRW, Iverson NR (2019) Genesis of glacial flutes inferred from observations at Múlajökull, Iceland. Geology 47(5):387–390. https://doi.org/10.1130/G45714.1

Leser H (1977) Feld- und Labormethoden der Geomorphologie. Walter de Gruyter, Berlin/New York

Penck A, Brückner E (1901–1909) Die Alpen Im Eiszeitalter, vol 3. Tauchnitz, Leipzig

Reichelt G (1961) Über Schotterformen und Rundungsgradanalyse als Feldmethode. Petermanns Geogr Mitt 105:15–24

Röthlisberger F, Schneebeli W (1979) Genesis of lateral moraine complexes, demonstrated by fossil soils and trunks; indicators of postglacial climatic fluctuations, paper presented at INQUA symposium on genesis and lithology of quaternary deposits, Int. Union for Quat. Res, Zurich

Rögner K (2020) Glazial geprägte Landschaften. In: Gebhardt H, Glaser R, Radtke U, Vött A (eds) Geographie. Springer Spektrum, Berlin, pp S 430–S 433

Supplementary Information

Glossary

Ablation Total of the processes by which a glacier loses mass. In addition to melting, this also includes sublimation, wind drift, → calving and → ice avalanches

Ablation area Area below the → equilibrium line on which the glacier records a mass loss at the end of the → budget year

Ablation period Season which is characterized by mass losses. In the → fixed date system, it lasts from 1st May to 30th September in the Alps

Abrasion Abrasive effect of glaciers on the bedrock; depending on the hardness of the abrasive, a distinction is made between → polishing and → striation

Accumulation area Area above the → equilibrium line on which the glacier records a mass gain at the end of the → budget year

Accumulation area ratio (AAR) Proportion of the → accumulation area to the total area of a glacier

Accumulation period Season which is largely determined by mass gains. In the → fixed date system, it lasts from 1st October to 30th April in the Alps

Accumulation Total of processes that add mass to the glacier. In addition to solid precipitation and resublimation, these include wind and avalanche input as well as refreezing of rain and meltwater

Anisotropy Directional dependence of a physical property or process

Arête Sharp mountain ridge, separating two cirques or glacial valleys and formed by glacial erosion

Basal sliding En-bloc movement of warm-based glaciers on the meltwater film between ice and → glacier bed

Basal transport zone Bottom meters of the glacier that may contain rock fragments from the → glacier bed

Bergschrund Glacier crevasse separating the uppermost ice frozen to the rock from the moving part of the glacier

Boulder clay Boulder-rich moraine material

Budget year Reference period for the annual → glacier mass balance. The natural budget year extends from an annual mass minimum (end of the melting period)

to the minimum in the following year and can vary calendrically due to different weather conditions. In the → fixed date system, the period is set (1 October to 30 September)

Bulldozing → Pushing

Calving Breaking of ice chunks or icebergs at → glacier fronts that flow into lakes or the sea

Chattermarks Transitional forms between → striae and → parabolic cracks caused by jerky yet continuous ice movement

Cirque (corrie, cwm) Bowl-shaped depression, open to the downhill side, located below the steepest section of a mountain and formed from a → nivation niche by glacial erosion

Cirque glacier Type of glacier that is spatially restricted to a → cirque

Cold glacier Glacier consisting exclusively of → cold ice

Cold ice Ice that is clearly below the → pressure melting point

Compensation effect Balancing effect of glaciers on the water flow in rivers, which results from the fact that ice melt is particularly high in those dry, hot summer periods when precipitation is absent

Compressive flow Compression processes in glacier ice when the flow velocity decreases downstream, for example at the bottom of a steep section

Confluence Junction of glaciers or glacial branches

Crag-and-tail Asymmetrical hill formed by glacial activity, consisting of a rock core with a steep stoss side (crag) and gently sloping softer material on the leeward side (tail)

Cryoconite Fine dust blown on glaciers by wind, which is glued together by micro-organisms to form granules

Cryoconite hole Hole in the ice surface caused by melting of small stones or → cryoconite accumulations

Dead ice Unmoving ice, separated from the active glacier

Dendritic glacier system → Ice field

Diagenesis Densification of snow and firn by overburden pressure

Diffluence Division of glaciers into different arms or valleys

Diffuse radiation The part of global radiation that is scattered in the atmosphere

Direct radiation The part of global radiation that reaches the earth's surface from the sun by the shortest route without being scattered in the atmosphere

Drumlin Ice-motion parallel, elongated and streamlined hill with a stoss face that is steeper than the lee face

Dry snow zone Upper hydrological zone on glaciers where melting never occurs. Exists only on very high or polar glaciers

Dumping Lateral sliding of → supraglacial rock material towards the glacier margin

Dynamic/volume response time Period of time it takes for a glacier to adjust to a sudden climate change and to find a new equilibrium with adjusted size and geometry

Early Holocene First stage in the → Holocene, which was characterized by striking glacial advances

Emergence Relative movement of the ice below the → equilibrium line directed towards the glacier surface

End moraine Moraine deposition at the glacier front by → pushing of unfrozen (→ push moraine) or by → thrusting of frozen loose material (→ thrust moraine)

Englacial Situated within a glacier

Equilibrium line altitude (ELA) Local snow line on glaciers. As the zero line of the mass balance, it separates the → accumulation area from the → ablation area

Erratic Large block of rock that was transported to its present position by glaciers, sometimes over several hundred kilometres, during a → glacial period

Esker Backfill of the central, subglacial drainage channel, which forms a dam-like wall (single ridge esker) in the direction of flow or may also be branched (braided esker)

Extending flow Stretching processes at locations where the flow velocity increases downstream, for example at the start of a steep section

False ogives Layer boundaries that are visible as outcrops due to tilting by ice movement. Seen from above, are bent downvalley by the faster ice movement in the middle of the valley. They have this shape in common with true → ogives, but unlike the latter they are not compressional ridges

Firn Snow, which has endured a whole summer and has a large density spectrum of 0.4–0.83 g cm^{-3}

Fixed date system Use of a fixed budget year from 1st October to 30th September in determining the → glacier mass balance

Fjord → Trough valley that empties into the sea and whose valley floor is flooded by the sea

Fluting (fluted moraine, glacial flute) Elongated ridge in the ground moraine that is formed by the permanent pressing of water-saturated fine material into → subglacial cavities and extends in the direction of flow

Foliation Layering in → glacier ice due to differences in crystal size and/or bubble content, caused by → shear stress and compression or representing sedimentary features

Fossil wood Decayed tree remains in → moraines that can help in dating the age of deposits and associated glacial advances

Friction cracks Crescent-like fractures in the bedrock that occur during jerky ice movement when a single particle of basal debris exerts high punctual pressure on the subsurface

Frontal Situated at the → glacier front

Furrow Channel-shaped, subglacial, glaciofluvial erosion form (P-form) in flow direction of the glacier

Geodetic method Method of mass balance determination based on height comparison of the glacier surface at two points in time. Mass changes are calculated from height differences, assuming an average density

Glacial geomorphology Subfield of landform science (Geomorphology) that deals with the processes and forms of glacial erosion and deposition

Glacial lake outburst flood (GLOF) Outburst of an ice- or moraine-dammed lake or water accumulation on the glacier (→ supraglacial lake) or in the glacier (→ water pocket)

Glacial milk Glacial meltwater with a high content of suspended matter. Due to the rock flour finely ground by the glacier, a light grey colour is characteristic of glacial streams until far downstream

Glacial mill Place where a → supraglacial meltwater stream disappears into the glacier interior. Glacial mills often form at crevasses or intersections of two crevasses

Glacial period (glaciation, glacial) Cool phase within an ice age in which large areas of the continental lands are glacierized

Glacial Concerning the glacier

Glacial series Model representation of a regular sequence of landforms formed by the activity of glaciers and their meltwaters at the margin of Pleistocene ice sheets and piedmont glaciers

Glacier bed Subsurface below the glacier, which may consist of solid or loose rock

Glacier foreland/forefield → Proglacial area in front of the glacier, which is usually not ice-free for very long

Glacier front Lower, i.e. valley-side end of the glacier

Glacier gate Tunnel opening at the valley-side end of the glacier, where the subglacial drainage channel leaves the glacier and becomes a glacial stream

Glacier ice Granular, air-impermeable end product of snow metamorphism with a density of 0.83–0.917 g cm^{-3}

Glacier mass balance Mass change of a glacier resulting from the difference between mass gain (→ accumulation) and mass loss (→ ablation). It relates to a specific period of time and is expressed as water equivalent (w.e.), i.e. as a water column in mm, cm or m

Glacier table Large stone or boulder that grows out of the glacier surface on an ice pedestal, protecting the ice shelf from solar radiation

Glacier tongue Elongated, lower part of the glacier that follows the course of the valley in its length

Glacierization Portion of a reference area (e.g. hydrological catchment, country, island) which is covered by glaciers

Glaciofluvial Formed by glacial meltwater

Glaciological method Direct method of mass balance determination in which the mass balance is measured on the glacier; the accumulation at snow pits and the ablation by stakes drilled into the ice

Glaciology The scientific study of the formation, properties, and nature of recent glaciers

Glacitectonics Deformation of unconsolidated, frozen sediments by advancing → glacier fronts. Compression and shortening in the → glacier foreland causes folding and thrusting

Global dimming Reduction of atmospheric transparency and thus solar radiation due to heavy air pollution, which led to cooling and glacier advances from the 1960s to the 1980s, especially in Europe

Global radiation Short-wave solar radiation arriving at the earth's surface, consisting of → direct radiation and → diffuse radiation

Grain shape The shape of rock particles, in geomorphology usually restricted to the aspect of rounding and subdivided into the four classes angular, sub-angular, sub-rounded and rounded

Grain size Diameter of a rock particle that can be assigned to the size classes boulders, cobble, gravel, sand, silt and clay

Gravel plain Glaciofluvial depositional surface analogous to a → sandur or ouwash plain, but consisting of coarser material (gravel)

Gravimetric method Method that determines → glacier mass balances from changes in the Earth's gravitational field

Ground moraine Subglacially formed glacial deposit without a conspicuous morphological form and with a high content of silt and clay

Hanging glacier Type of glacier on steep mountain slopes where → ice avalanches are the dominant process of → ablation

Hanging valley Side valley whose valley floor lies higher than that of the main valley, because it has been less eroded due to a lower ice thickness

Holocene climatic optimum Warmest phase in the → Holocene 7000–6600 years ago

Holocene Period from the end of the last → glacial period (11,700 years before present) to the present

Horn Pyramid-like mountain peak, sharpened on all sides by glacial erosion

Hydrological-meteorological method Method that calculates the mass balance of a glacier as a residual term of the water balance equation (precipitation minus evaporation minus runoff)

Ice Age Relatively cool period in Earth's history in which at least one pole is glacierized

Ice avalanche A mass of ice that has detached from a glacier and is falling downhill. Part of the ablation process, particularly at hanging glaciers

Ice cap Very large (up to 50,000 km²) glacier in polar regions, overlying distinct relief

Ice cliff Exposed ice flank on → debris-covered glaciers, emerging on spots where the → supraglacial moraine has slipped off

Ice field Type of glacierization in which mountain glaciers in different valleys are connected to each other via passes. Exists today only in polar regions

Ice sheet Glacier of continental extent, existing today only in Greenland and Antarctica

Ice-dammed lake Glacial lake whose dam is made of glacier ice

Icefall Steep section in the longitudinal profile of a glacier where the ice breaks up through transverse crevasses and can dissolve into individual → séracs

Inland ice → Ice sheet

Inner moraine Rock material within glacier ice (→ englacial debris)

Interglacial period (interglacial) Warm phase within an ice age in which the glaciers are comparatively small

Internal deformation Main form of ice movement by creep

Intraglacial → Englacial

Jökulhlaup Special form of → glacial lake outburst floods, translated from Icelandic as "glacier run". In volcanic areas, geothermal or eruptive activity can lead to large subglacial water accumulations that grow to a critical volume and then erupt. In the English-language literature, the term is often used more broadly and generally for glacial lake outbursts

Kame Deposition of material on or in glaciers or between dead ice blocks, which remain after the melting of the ice. Surface deposits from supraglacial lakes are called kame plateaus; accumulations on the valley slope are called kame terraces

Kame-and-kettle topography Chaotic topography formed by thawing ice masses and debris-laden meltwater streams, in which → kettles are associated with → kames

Karakoram anomaly Phenomenon of currently stable or even slightly advancing glaciers in the Karakoram and adjacent mountain ranges, probably due to temporarily increased snowfall

Kettle hole Depression created when a block of → dead ice prevents a site from being filled by (glacio-)fluvial sediments. The hollow is often seasonally wet (kettle pond) or permanently water-filled (kettle lake)

Kettle lake → Kettle hole

Kettle pond → Kettle hole

Last Glacial Maximum (LGM) Phase of the largest ice advances during the youngest →glacial period

Late glacial Period between the → Last Glacial Maximum (approx. 20,000 years before present) and the end of the Würm glaciation (11,700 years before present), characterised by ice disintegration and retreating of the glaciers back to modern dimensions

Latent heat flux Energy flux through phase transitions of water

Lateral Situated at the side of a glacier

Lateral moraine Glacial deposit formed by → dumping at → lateral glacier margins

Latero-frontal Located at the transitional area between → lateral and → frontal

Lichenometry Method for dating moraines based on the uniform growth of lichens

Little Ice Age (LIA) Coldest phase of the → Holocene from approx. 1450 AD to 1850 AD

Lodgement Active accumulation under moving ice

Lodgement till Moraine deposit formed by → lodgement at the → glacier bed

Longitudinal crevasse Crack formed by lateral expansion when valleys are widened or at the margin of mountains

Longwave radiation Electromagnetic energy emitted from the earth's surface, strongly dependent on surface temperature

Lunate fracture Coarser, crescent-shaped → friction crack in bedrock, the open sides of which face away from the glacier

Marginal crevasse Glacial crevasse formed by shear stress due to velocity decrease from the central flow line towards the margin

Marine isotope stage (MIS) Warm or cold period in the history of the Earth, inferred from the oxygen-isotope ratio in marine sediments

Medial moraine Elongated debris ridge on the glacier, in flow direction, formed by the junction of two → lateral moraines at the → confluence of glaciers or by large, punctual accumulations of basal debris in conjunction with → emergent ice movement

Medieval climate optimum Warm phase in the Holocene from 750 AD to 1500 AD

Melt-out Passive melt-out when ice is very slow or stagnant

Melt-out till Moraine deposit formed by → melt-out at the → glacier bed (subglacial melt-out till) or the glacier surface (→ supraglacial melt-out till)

Metamorphosis (of snow) Transformation process in which → new snow is compacted via → old snow to → firn and finally → glacier ice. In general, a distinction is made between → destructive and → constructive metamorphosis

Destructive metamorphosis Transformation process in which snow crystals become smaller and smaller. This happens at low temperature gradients in the snow pack (isothermal metamorphosis) either mechanically by pressure or via the gas phase. Another form of destructive metamorphosis is the transformation of snow grains by melting and refreezing (melt-freeze metamorphosis)

Constructive metamorphosis Transformation process at high temperature gradients in the snow pack, where water vapour diffusion from warmer to colder crystals results in the formation and enlargement of snow grains. As the transformation progresses, the snow cover deconsolidates and cup crystals (→ depth hoar) are formed

Migration Period Pessimum Cold phase in the → Holocene from 450 AD to 700 AD

Moraine Landform composed of glacial till, deposited by a glacier

Moraine-dammed lake Glacial lake whose dam is made of moraine material

Moulin-kame Former filling of a → glacial mill, forming a mound after melting

Muschelbruch Crescent-shaped, subglacial, glaciofluvial erosion form (P-form) transverse to the flow direction of the glacier

Neptunist A follower of a geological doctrine after Abraham Gottlob Werner (1749–1817), according to which rocks are formed by sedimentation processes in the ocean

Net balance → Glacier mass balance

Net radiation Radiation exchange of a system with its environment, calculated as the difference between the radiation fluxes to and from the system (here: glacier)

Net-adfreezing Local freezing of debris at the base of the glacier

New snow Freshly fallen snow with a density below 0.2 g cm^{-3} and an air content of more than 90%

Nivation niche Depression, usually at the base of steeper rock faces, enlarged by snow creep and sliding, melt water runoff and enhanced chemical weathering

Nunatakk Mountain peak or rock island that protrudes isolated from a glacier. Typical nunatakker are only found in strogly glacierized regions, e.g. at ice fields, ice caps and at the marginal areas of ice sheets

Nye channel (N-channel) Subglacial drainage channel incised into the glacier bed

Ogive Bulge in the glacier surface below an → icefall, where faster ice meets slower ice (→ compressive flow). The bulge becomes the typical arc shape, because ice velocity is highest along the central flowline in the middle of the glacier

Old snow Slightly metamorphically altered snow with a density of 0.2–0.4 g cm^{-3}

Optimum Phase in the Holocene when the climate was warmer than today

Outlet glacier → Valley glacier flowing off → ice caps or → inland ice

Outwash plain → Sandur

Parabelriss Finer, crescent-shaped friction crack in the bedrock whose open sides face up-glacier

Percolation zone Hydrological zone on glaciers in which meltwater seeps from the snow surface into deeper layers and refreezes there. However, a layer of dry winter snow still exists below the layers where refreezing occurs

Permafrost Permanent freezing in solid rock or loose rock that lasts for at least 2 years

Pessimum Phase of colder climate in the Holocene, mostly evidenced by glacial advances

P-form (plastically moulded form) Erosion form created below the ice by flowing water under high isostatic pressure

Piedmont glacier Type of glacier that develops when a glacier leaves the laterally constricting relief of a mountain range in order to then flow out in all directions in the foothills, which have little relief

Pleistocene Ice Age, which began 2.6 million years ago and ended 11,700 years ago. It contains an as yet unknown number of → glacials and → interglacials

Plucking Removal of fragments from heavily fractured sections in the lee of rock obstacles

Plutonists A follower of a geological doctrine after James Hutton (1726–1797), according to which rocks are formed by magmatic processes in the depth

Polishing Variant of → abrasion in which the removal of rock protrusions in the micro-range results in a smoothing of the surface (glacial polish). Polishing occurs primarily when the eroding particles have the same hardness as the bedrock

Polythermal glacier Glacier that has areas of both cold and warm ice

Pothole Circular cavity in bedrock formed by water vortices in crevasses and → glacial mills and belonging to the non-directional → P-forms

Pressure melting point Pressure-dependent melting point, which decreases by 0.0073 °C per bar increase in pressure

Proglacial Situated in front of the glacier, i.e. from the glacier front down the valley

Progressive erosion Outbreak mechanism in moraine-dammed lakes, in which the lowering of the outflow level causes an increase in outflow discharge and thus results in a self-reinforcement

Push moraine Proglacial moraine, caused by → bulldozing of unfrozen debris by advancing glaciers

Pushing Formation of → push moraines by advancing glacier front compressing unfrozen debris

Radiocarbon method (^{14}C method) Dating method for organic material based on the radioactive decay of ^{14}C and applicable to biomass up to 60,000 years old

Rat tail Elongated, streamlined elevation in solid rock that forms in the protection of a more resistant rock section on the lee side of the same by selective abrasion

Reconstituted glacier Glacier that is interrupted by a steep bedrock section. The lower part is fed by → ice avalanches that fall from the upper part

Riegel Elevation in the longitudinal profile of a valley floor that occurs at locations where the rock is more resistant to erosion or at → diffluences, where erosion rates decrease

Roches moutonnés Glacially shaped rock obstacle with a flat up-glacier side due to → abrasion and steep down-glacier side due to → plucking

Rock drumlin Glacially shaped rock obstacle deviating from the standard → roche moutonnée shape by showing a steep up-glacier and flat down-glacier side. Below slowly moving ice, → abrasion at the stoss side is ineffective. At the lee side, no subglacial cavity is formed, resulting in the absence of → plucking

Roman climatic optimum Warm phase from 300 B.C. to 400 A.D.

Röthlisberger channel (R-channel) Meltwater channel that is → englacial or → subglacial. The subglacial variant is completely formed in the ice and, in contrast to the → Nye channel, is not incised into the glacier bed

Sensible heat flux Direct heat transport, which manifests itself through a change in temperature

Sérac Ice column, often formed by widely opened, intersecting crevasses in an icefall

Shear stress Component of stress coplanar with a material cross section. Shear stress deforms, for example, a rectangular body into a parallelogram

Sichelbruch Coarser, crescent-shaped friction crack in the bedrock whose open sides face up-glacier

Sichelwanne Crescent-shaped, subglacial, glaciofluvial erosion form (P-form) transverse to the flow direction of the glacier

Slush zone Water-saturated layer in the wet snow zone, where sudden outflow, so-called slush flows, can occur

Snow line Altitude above which snow exists permanently

Climatic snow line Value averaged over several years and over a region, which applies to an entire mountain group

Orographic snow line Local value averaged over several years. Can take on different values on sunny and shady sides of the same mountain

Temporary snow line Current and local value. Lower limit of the current snow distribution, is subject to large seasonal fluctuations

Snow penitents Jagged snow surface formed by differential → ablation of snow

Snowline depression Drop in the equilibrium line that was necessary to enable a certain glacier advance. The reference level used is usually not today's equilibrium line altitude, but that of the Little Ice Age maximum. Snow line depressions are used to correlate glacier advances between different valleys and regions

Sorting Scattering of the grain sizes of a sediment; the stronger the sorting, the lower the scattering and the more uniform the size of the individual particles

Splaying crevasses Glacier crevasses formed by lateral extension with unobstructed flow to all sides, e.g. on → piedmont glaciers

Stadial Intermediate advance within the general trend of glacier retreat during the → Late glacial

Stratification The succession of layers, or homogeneous deposits, in sediments

Striae → Striation

Striation Variant of abrasion in which debris in the glacier bed causes scratches in the bedrock. Both the process and the result are called striation (plural: striae). Striation is most effective when the debris is significantly harder than the glacier bed

Sub-debris ablation Melt below a debris layer

Subglacial Situated beneath the glacier

Submergence Movement of snow, firn and glacier ice above the → equilibrium line directed into the glacier interior

Summer accumulation type glacier Glacier in which the maxima of mass gains and mass losses occur simultaneously during the warm season. These conditions mainly occur on tropical and monsoon-influenced glaciers

Superimposed ice zone Hydrological zone on the glacier surface which consists of ice that has been formed by refreezing of water-saturated snow. It can originate at the surface or at the bottom of the snowpack, in the latter case coming to the surface by subsequent melting of the overlying snow

Supraglacial Situated on the glacier surface

Surge Rapid advance of a glacier due to the release of a mass imbalance that was built up over several years or decades. Surges are not caused by climate fluctuations

Tarn Water-filled → cirque

Temperate glacier Glacier consisting exclusively of → warm ice

Temperate ice → Warm ice

Terminal moraine → End moraine

Terminus response time Time delay from the climate signal to the response of the glacier front

Thrust moraine Terminal moraine formed by → thrusting

Thrusting Deformation of frozen debris by advancing glaciers, whereby shortening causes folding and shearing (→ glacitectonics) and → thrust moraines are formed

Till Material of a glacial deposit that is characterized by certain physical properties (grain size distribution, porosity, density, etc.). From a sedimentological point of view, till is generally angular, unstratified and unsorted

Tongue basin Hollow formed by erosion of piedmont glaciers, often filled with water

Transverse crevasse Glacier crevasse transverse to the direction of flow, which usually occurs where the terrain suddenly becomes steeper and the the flow verlocity increase abruptly (→ extending flow)

Trough U-shaped valley cross section created by glacial erosion

Unconstrained glacier Type of glacier that covers the bedrock so completely that the flow direction of the ice is decoupled from it and is only controlled by the inclination of the ice surface

Valley glacier Mountain glacier with a distinct glacier tongue

Valley train → Sandur in mountain relief, which is limited in its lateral extent

Warm ice Glacier ice that is at the pressure melting point

Water pocket A collection of water in the glacier that can empty spontaneously and is a potential natural hazard

Wet snow zone Hydrological zone below the percolation zone, where the entire snow pack is affected by melting and refreezing

Whaleback Glacially formed rock obstacle that deviates from the standard → roche moutonnée shape by being symmetrical. At moderate flow velocities, → abrasion on the stoss side increases compared to the → rock drumlin, but a subglacial cavity with → plucking on the lee side cannot yet develop

Winter accumulation type glacier Glacier in which mass gain occurs in winter, seasonally separated from the summer ablation period. Most glaciers in the middle latitudes belong to this type

Index

Printed in the United States
by Baker & Taylor Publisher Services